科技英语阅读教程

主　编　尹丕安
编　者　（按姓氏笔画排序）
王云鹤　付一凡　闫　敏　闫维雪
张凯凯　金玉凡　姜　震　袁　洁

科学出版社
北　京

内 容 简 介

本书总共8个单元，本着专业性和知识性相结合的编写原则，重点在于培养学习者专门用途英语的阅读能力，兼顾写作能力的培养。本书涵盖内容广泛，包括商贸、人文艺术、信息技术、航空、材料、机械工程、经管、金融会计等方面。每个单元分为A和B两篇文章，A篇强调语言应用技能、阅读技巧和写作能力的培养。B篇注重学习者专业知识的拓展、跨文化科技交流能力和科技英语词汇量的提升。

本书既可用于非英语专业本科生和研究生的科技英语教学，也可供对科技英语感兴趣的自学者使用。

图书在版编目（CIP）数据

科技英语阅读教程 / 尹丕安主编. —北京：科学出版社，2020.11
ISBN 978-7-03-066383-2

Ⅰ. ①科…　Ⅱ. ①尹…　Ⅲ. ①科学技术–英语–阅读教学–高等学校–教材　Ⅳ. ①N43

中国版本图书馆CIP数据核字（2020）第198971号

责任编辑：常春娥 / 责任校对：贾伟娟
责任印制：李　彤 / 封面设计：蓝正设计

科学出版社 出版
北京东黄城根北街16号
邮政编码：100717
http://www.sciencep.com
北京盛通商印快线网络科技有限公司 印刷
科学出版社发行　各地新华书店经销
*
2020年11月第 一 版　开本：787×1092　1/16
2021年 1 月第二次印刷　印张：7 1/4
字数：300 000

定价：98.00元

（如有印装质量问题，我社负责调换）

前　言

“科技英语”教学是我国大学本科公共英语教学的一个有机组成部分，是基础英语学习到专业领域英语学习之间的一个过渡，是对大学公共英语教学的一个扩展和补充。经过长期的教学实践，我国大学英语教学已经比较成熟，教学模式和教材建设也相对稳定下来。但是，由于“科技英语”不是必修和主干课程，针对“科技英语”的教材建设相对比较滞后，无法满足大学英语教学的需求。许多非英语专业类学生，尤其是理工类的学生，科技英语阅读能力薄弱，在阅读和理解英语原版科技类著作和资料时，遇到不少困难，因而不能透彻理解原文。同时，很多理工类学生的科技英语写作能力也比较薄弱，不能够熟练掌握运用科技英语的语法特点，本教材正是基于这个大背景而编写的。

21 世纪科学与技术飞速发展，英语已成为全球范围内科技人员进行科学研究和学术交流必不可少的工具。科技英语的掌握程度在科学技术人员查询与阅读科技文献、了解科技前沿发展状况、参加国内外学术交流等方面，起着越来越大的作用。随着我国科技实力进一步提升，对外开放全方位进行，中国科技工作者在学术领域的对外交流越来越频繁，在科学与技术领域与西方同行相互交流的机会越来越多，因此科技英语的重要性愈发突显，提高科技人员科技英语阅读能力和科技交流能力的需求愈来愈紧迫。

大学“科技英语”教学是本科生培养的一个重要组成部分，其旨在帮助学生完成从大学基础英语阅读阶段到专业英语阅读阶段的过渡。其教学目标应该具有两方面：一是培养学生的国际化视野；二是提高学生专业学术英语能力。科技英语语篇具有丰富的科技词汇、独特的语法结构，以及专业上通用的表达方式。学习科技英语是对大学基础英语学习的补充和提高，也是学习者开阔视野、理解世界范围内专业前沿科学技术和知识发展现状的必要途径。通过本教材的学习，学习者可以了解科技英语的句式表达方式特点、行文方法在学术英语中的具体表现，为以后阅读专业英语文献和英文原著打下坚实的基础。同时，学习者可以进一步提高英语阅读理解和综合分析能力以及信息转换能力，在阅读真实的科技语言材料的基础上，扩大自己的科技词汇量，开阔科普视野和思路。针对大学本科生来源多元化的现状，一方面要注重提升其跨文化的学术交流能力以及语言知识和语言技能的应用能力，同时还要兼顾培养学生运用外语解决专业问题的能力，使外语学习能更好地为其专业学习和研究打好基础，最终提高他们学习和工作的竞争力。

基于以上教学目标和原则，笔者专门编写了这本教材。本教材的编写主要应用于大学非英语专业高年级学生的选修课程。本教材在内容上选用了商贸、人文艺术、信息技术、航空、材料、机械工程、经管、金融会计等几大类的内容，共 8 个单元，每个单元由 2 篇文章构成。课后练习部分分为词汇练习、语法练习、翻译练习和写作练习几大部分。本教材的文章均选自国内外英文著作、英文报纸、英文期刊等对个别内容进行了修正。文章时效性强，语言难度适中，地道实用，适合非英语专业本科生和研究生的课堂教学和自主学习。

本教材的编写特点是：第一，自始至终体现科技英语的词汇和语法特点。科技英语的词汇及其解释不仅出现在词汇表中，在每篇课文的练习当中，也都针对性地突出每篇课文的重点科技词汇和句型的使用，以使学习者在认知层面加深对科技英语词汇和语法特点的掌握。第二，科技阅读材料多元化，同时注重学习者逻辑思维、批判思维的培养和知识面的扩充。所选材料既考虑到最新科技的发展，也兼顾到人文科学方面。第三，兼顾学习者科技阅读能力和科技写作及翻译能力的提升。通过正文的科技阅读训练，学习者在课后的练习中也可以着重加强自己的科技翻译和写作能力训练。

书中纰漏之处在所难免，欢迎读者提出宝贵意见和建议。

尹丕安

2020 年 5 月

CONTENTS

Unit 1 Business and Trade

✧ Text A

Business and Your Life

You Depend on Business

Modern business is well organized and operates smoothly. We often accept business services, therefore, without much thought of their importance to us. Have you ever considered the many ways in which you depend on business? Business supplies the food you eat, the clothes you wear, the home you live in, and the many other goods and services used in satisfying your wants.

Many times a week most of us are affected by business. During a certain week, for example, you may make telephone calls, have your eyeglasses repaired, ride in a taxi, consult your doctor, deposit money in your savings account, buy a new bicycle tire, or advertise for your lost dog.

Your home, too, depends on business in obtaining the goods and services your family needs. One family reported these business activities for a certain day: issued checks to pay automobile insurance and electric bill; bought a rug and chairs; mailed a letter to order some books; paid the newsboy; hired a carpenter to repair a window frame; bought a United States savings bond at the bank; left a roll of film at the camera shop; had dinner at a restaurant.

What Is Business?

You frequently hear the word business in everyday conversations. Here are some expressions using the word: "How is business this month?" "What line of business is he in?" "Our firm does a cash business." "Business hours are 9 to 5." "John is taking a business course." Although you may have a general understanding of the meaning of each statement, at this time you probably would have difficulty in giving a clear-cut explanation of business.

Business means different things to different people. To one person, it means producing goods through farming, manufacturing, or some other industry. To another, it means buying and selling merchandise. To a third person, it means providing services. To a fourth, it means engaging in an occupation to earn a living. Each of these instances—making goods, buying and selling goods, providing services, engaging in an occupation—illustrates the meaning of business. In this article, business means the work or activities by which goods and services are provided and obtained for money payment.

Some Activities Are Not Business

Enterprises such as the grocery store, the bus line, the newspaper, the electric company, and the bank are readily identified as being in business. They supply goods and services for payment. What about the laundry, the television repairman, the plumber? They, too, work for payment and are therefore engaged in business.

Not all activities in which work is involved are classed as business. If you help a friend repair a car or paint his house, you would be working. Unless you receive payment, however, you are not taking part in business in the true sense of the word. Here is the test of whether or not an activity can be classed as business: Is payment made for the goods supplied or service performed? If money payment is required, the activity is business.

Business Transactions

Producer and consumer. To carry on business, whether it consists of producing goods, distributing goods, or providing services, three factors—people, goods, and money—are necessary.

People refer to both producers and consumers. Producers are the firms and workers who produce and distribute goods or services. Consumers are those who buy and use goods or services.

Individual wants and community wants. Goods mean the products that persons or communities buy to satisfy their wants. Individual wants consist of the necessities, comforts, and luxuries of life. Goods that everyone must have, such as food, clothes, and shelter, are called necessities. Goods and services, such as books, telephones, electricity, and gas, which make life easier and more enjoyable, are called comforts. Nonessential high-quality goods, such as expensive jewelry, yachts, and customer-built cars, are called luxuries.

The needs that arise when people live in groups are referred to as community wants. Communities must have police and fire protection, water system, highways, stores, schools, and banks.

Money is exchanged for goods and services. Many years ago trade was carried on by barter, which is exchanging one article for another—for example, trading a knife for a pair of shoes. Today we usually pay money for what we want. The exchange of money for goods or services is a business transaction.

We need business. Business provides us with the things we use—food, clothes, and many other products and services. Any interruption of essential activities in your community would create much hardship. Few families keep on hand enough food and other essentials for more than a brief period, and any stoppage in the flow of products would result in considerable suffering. Imagine the situation if no railroads or trucks were operating; if there were no deliveries of bread, milk, or other goods; no electricity and gas; no store, bank, factory, or restaurant open!

If just one phase of business, such as transportation, were to suspend operations, factories could not ship their products. Soon workers would be laid off because of shutdowns; and with the loss in wages, families would have to curtail their buying. The resulting chain of events could easily bring on a condition of business paralysis. Only when the business activities of the community, the state, and the nation are functioning smoothly can there be prosperity for everyone.

For the consumer, business supplies the goods and services he must have to live and take care of his affairs properly. For the worker, business provides employment and therefore a means of earning a living. For the business owner business activities offer the opportunity to render a service and to make a profit. For the investor, business provides ways to put his funds to work.

Because it furnishes the things we use, gives us useful work to do, offers opportunities for saving and investing, and aids the national defense, business is truly the backbone of modern life. In fact, business is the foundation on which the existence and welfare of our people depend.

(Adapted from *Business and Your Life*. https://www.doc88.com/p-1425909841423.html [2019-6-17])

☞New Words

1. transaction [træn'zækʃn] *n.* a piece of business, for example, an act of buying or selling something 交易
2. deposit [dɪ'pɒzɪt] *n.* to put money into a bank account 将（钱）存入银行；存储
3. laundry ['lɔ:ndri] *n.* a business or place where you send sheets, clothes, etc. to be washed 洗衣房
4. yacht [jɒt] *n.* a large boat with sails or a motor, used for racing or pleasure trips 快艇
5. suspend [sə'spend] *v.* to delay it or stop it from happening for a while or until a decision is made about it 暂停
6. carpenter ['kɑ:pəntə(r)] *n.* a person whose job is making and repairing wooden things 木匠
7. clear-cut [ˌklɪə'kʌt] *adj.* easy to recognize and quite distinct 清晰的
8. instance ['ɪnstəns] *n.* a particular example or occurrence of something 例子
9. barter ['bɑ:tə(r)] *v.* to exchange goods, property, services, etc. for other goods, etc. without using money 以物交换；以物易物
10. shutdown ['ʃʌtdaʊn] *n.* the closing of a factory, shop, or other business, either for a short time or forever 歇业

11. curtail [kɜːˈteɪl] *v.* to limit sth. or make it last for a shorter time 限制，缩短
12. consumer [kənˈsjuːmə(r)] *n.* a person who uses goods or services 消费者
13. render [ˈrendə(r)] *v.* to give sb. sth. especially in return for sth. or because it is expected 给予，提供
14. furnish [ˈfɜːnɪʃ] *v.* to provide or supply sb./sth. with sth. 向（向某人或某物）供应
15. backbone [ˈbækbəʊn] *n.* the most important part of a system, etc. that gives it support and strength 支柱
16. paralysis [pəˈræləsɪs] *n.* the state of being unable to act or function properly 瘫痪状态
17. prosperity [prɒˈsperəti] *n.* a condition in which a person or community is doing well financially 繁荣
18. automobile [ˈɔːtəməbiːl] *n.* car 汽车
19. insurance [ɪnˈʃʊərəns] *n.* an arrangement in which you pay money to a company, and they pay you if something unpleasant happens to you 保险
20. film [fɪlm] *n.* the narrow roll of plastic that is used in a camera to take photographs 胶卷
21. conversation [ˌkɒnvəˈseɪʃn] *n.* the use of speech for informal exchange of views or ideas or information etc. 交谈
22. jewelry [ˈdʒuːəlri] *n.* an adornment (as a bracelet or ring or necklace) made of precious metals and set with gems (or imitation gems) 珠宝
23. enterprise [ˈentəpraɪz] *n.* a company or business 公司；企业
24. plumber [ˈplʌmə(r)] *n.* a craftsman who installs and repairs pipes and fixtures and appliances 水管工
25. individual [ˌɪndɪˈvɪdʒuəl] *n.* person 个人
26. necessity [nəˈsesəti] *n.* something that you must have in order to live properly or do something 必需品
27. luxury [ˈlʌkʃəri] *n.* something expensive which is not necessary but which gives you pleasure 奢侈品
28. arise [əˈraɪz] *v.* (especially of a problem or a difficult situation) to happen（尤指问题或困境）发生；出现
29. welfare [ˈwelfeə(r)] *n.* the general health, happiness and safety of a person, an animal or a group（个体或群体的）幸福，福祉，安康
30. foundation [faʊnˈdeɪʃn] *n.* the basis on which something is grounded 基础
31. community [kəˈmjuːnəti] *n.* all the people who live in a particular area or place 社区

☞Phrases and Expressions

1. earn a living: to keep the pot boiling; to make one's living 谋生
2. customer-built car: a car made at the customer's request 按买主要求制造的汽车
3. lay off: to discharge a worker 解雇

4. make a profit: to earn money 赚钱
5. consist of: to be constitutive of 由……组成；由……构成
6. depend on: to be contingent on 取决于……
7. exchange for: to turn/give for 交换

☞Notes

1. savings bond: non-negotiable government bond; cannot be bought and sold once the original purchase is made（由美国政府发行的）储蓄公债
2. savings account: an account at a bank that accumulates interest 银行账户

☞Exercises

Ⅰ. Reading Comprehension

Answer the following questions according to the text.

1. What is the best definition of "business" according to this text?
2. What activities belong to business? And what are not business activities?
3. What are the three essential elements to do business?
4. What would happen if one phase of business, such as transportation, were to suspend operations?
5. Why do we say that business is truly the backbone of our modern life?

Ⅱ. Vocabulary

A. Replace the underlined words or expressions in the following sentences with one of the best choices from the word bank. Change the form when necessary.

outset	bounce	put	derive	typical	interfere	schedule	develop	case
spread	plunge	use	appear	choose	different			

1. After 2 weeks' holiday, she <u>started feeling better or stronger</u> with renewed vigor.
2. The author may choose to use a characteristic work, something <u>characterized by</u> its time and place.
3. These differences may <u>lie in</u> their size, their history, their role, and the personalities of their managers.
4. Police say the thieves managed to <u>have a damaging effect on</u> the alarm system before breaking into the shop.
5. <u>Sometimes</u> there are no symptoms at all to indicate that one has a low iron reserve.

6. The industrial revolution which started in Britain reached other countries in Europe.
7. The new school is supposed to open at the end of August.
8. It is known to all that knowledge grows out of books and experience.
9. Mr. John suggested a five-point plan at the meeting for promoting sales.
10. We are utterly unprepared for the change of the weather. At the beginning, it looked like a nice day.

B. Fill in the blanks in each sentence by selecting the most suitable words.

1. He ____ interrupted me by asking irrelevant questions.
   A. continually B. continuously C. consistently D. consequently
2. Since ____ can't work in the United States without a permit, so it is of great importance for them to present their credentials to the government.
   A. emigrants B. expatriates C. migrants D. immigrants
3. Most investors are taught at the very beginning that there is no place for ____ in investment markets.
   A. feeling B. emotion C. passion D. sentiment
4. Lisa would rather ____ at home than ____ to the cinema .
   A. staying; going B. staying; go
   C. to stay; to go D. stay; go
5. Advertising is distinguished from other forms of communication ____ the advertiser pays for the message to be delivered.
   A. in that B. which C. whereas D. because
6. The ____ driver thinks accidents only happen to other people.
   A. normal B. usual C. average D. common
7. Our ____ sensitivity decreases with age. By age 60, most people have lost 40 percent of their ability to smell and 50 percent of their taste buds.
   A. sensible B. sensory C. sensitive D. senseless
8. More and more Chinese students find that their native culture will unavoidably ____ the foreign ones as they are making more contacts with the world.
   A. conform with B. stumble on C. collide with D. trifle with
9. Nowhere in nature is aluminum found free, owing to its always ____ with other elements, most commonly with oxygen.
   A. being combined B. having combined
   C. to combine D. combined
10. Physics is the present-day equivalent of ____ used to be called natural philosophy, from ____ most of present-day science arose.
   A. which, what B. that, which C. what, which D. what, that

Ⅲ. Cloze

Choose an appropriate word from the four choices marked A, B, C and D for each blank in the passage.

Faces, like fingerprints, are unique. Did you __1__ wonder how it is possible for us to recognize people? Even a skilled writer probably could not describe all the __2__ that make one face different from another. Yet a very young child—__3__ an animal, such as a pigeon—can learn to recognize faces. We all __4__ this ability for granted.

We also tell people apart by how they behave. When we talk about someone's personality, we mean __5__ in which he or she acts, speaks, thinks, and feels that __6__ an individual different from others.

Like the human face, human personality is very complex. But describing someone's personality in words is somewhat easier than __7__ his face. If you were asked to describe what a "nice face" looked like, you __8__ have a difficult time doing so. But if you were asked to describe a "nice person", you might begin to think about someone who was kind, __9__, friendly, warm, and so forth.

There are many words to describe how a person thinks, feels, and acts. Gordon Allport, a U.S. psychologist, found nearly 18,000 English words characterizing differences in people's behavior. And many of us use this information as a __10__ for describing, or typing, a personality. Hippies, bookworms, conservatives, military types—people are described with such terms.

1. A. sometimes	B. always	C. ever	D. anytime
2. A. features	B. distinctions	C. characteristics	D. qualities
3. A. or even	B. and then	C. and also	D. and too
4. A. have	B. take	C. use	D. regard
5. A. the manners	B. the means	C. the ways	D. the patterns
6. A. causes	B. makes	C. cause	D. make
7. A. describe	B. to describe	C. describing	D. description of
8. A. will	B. would	C. shall	D. should
9. A. considerate	B. considering	C. considerable	D. concerning
10. A. one	B. point	C. basis	D. criterion

Ⅳ. Translation

Translate the following text into Chinese.

Getting a proper amount of rest is absolutely essential for building your energy resources. If you frequently work far into the night or have poor sleep, it stands to reason that you may start to feel a little run-down. Though everybody is different, most people need at least seven to eight hours of sleep per night in order to function at their best.

If you have been lacking energy, try going to bed earlier at night. If you can wake up feeling well-rested, it will be an indication that you are starting to get an appropriate amount of sleep at night. If you sleep more than eight hours every night but still don't feel energetic, you may actually be getting too much sleep.

Once in a while, you are bound to have nights where you don't get an adequate amount of sleep. When your schedule permits, you can also consider taking a short sleep during the day, for sometimes taking a nap is the perfect way to recharge your batteries.

Ⅴ. Writing

Today, online shopping is becoming more common. Consumers can buy almost everything they need online. What do you think about online shopping? Write about the following topic with no less than 150 words. Give reasons for your answer and include any relevant examples from your own knowledge or experience.

Shopping on the Internet

✧ Text B

What Is International Trade?

International trade is the exchange of goods and services across international borders. In most countries, it represents a significant share of Gross Domestic Product (GDP). While international trade has been present throughout much of history, its economic, social, and political importance has been on the rise in recent centuries, mainly because of industrialization, advanced transportation, globalization, multinational corporations, and outsourcing. In fact, it is probably the increasing prevalence of international trade that is usually meant by the term "globalization".

Traditionally international trade was regulated through bilateral treaties between two nations. For centuries under the belief in Mercantilism, most nations had high tariffs and many restrictions on international trade. In the nineteenth century, especially in Britain, a belief in free trade became paramount and this view has dominated thinking among western nations most of the time since then. In the years since the Second World War multilateral treaties like the General Agreement on Tariffs and Trade (GATT) and World Trade Organization have attempted to create a globally regulated trade structure.

No nation can meet all of its people's needs and every state engages in at least some sort of international trade.

Free trade is usually most strongly supported by the most economically powerful nation in the world. The Netherlands and the United Kingdom were both strong advocates of free trade when they were on top; today the United States, the European Union and Japan are its greatest proponents. However, many other countries—including several rapidly developing nations such as India, China and Russia—are also becoming advocates of free trade.

Traditionally agricultural interests are usually in favor of free trade while manufacturing sectors often support protectionism. This has changed somewhat in recent years, however. In fact, agricultural lobbies, particularly in the United States, Europe and Japan, are chiefly responsible for particular rules in the major international trade treaties which allow for more protectionist measures in agriculture than for most other goods and services.

During recessions, there is often strong domestic pressure to increase tariffs to protect domestic industries. This occurred around the world during the Great Depression leading to a collapse in world trade that many believe seriously deepened the depression.

The regulation of international trade is done through the World Trade Organization at the global level, and through several other regional arrangements such as MERCOSUR in South America, NAFTA between the United States, Canada and Mexico, and the European Union between twenty-seven independent states. There is also the Free Trade Area of the Americas (FTAA), which provides common standards for almost all countries in the American continent.

As far as free trade is concerned, free-market economy plays the most important role. Economists describe a market economy as one where goods and services are exchanged at will and by mutual agreement. Buying vegetables for a set price from a grower at a farm stand is one example of economic exchange. Paying someone an hourly wage to run errands for you is another example of an exchange.

A pure market economy has no barriers to economic exchange: you can sell anything to anyone else for any price. In reality, this form of economics is rare. Sales taxes, tariffs on imports and exports, and legal prohibitions—such as the age restriction on liquor consumption—are all impediments to a truly free market exchange.

In general, capitalist economies, which most democracies like the United States adhere to, are the freest because ownership is in the hands of individuals rather than the state. Socialist economies, where the government may own some but not all the means of production can also be considered market economies as long as market consumption is not heavily regulated.

There's a reason why most of the world's most advanced nations adhere to a market-based economy. Despite their many flaws, these markets function better than other economic models. Here are some characteristic advantages and drawbacks:

Competition leads to innovation. As producers work to satisfy consumer demand, they also look for ways to gain an advantage over their competitors. This can occur by making the production process more efficient, such as robots on an assembly line that relieve workers of the most monotonous or dangerous tasks. It can also occur when a new technical innovation leads to new markets, much as when the television radically transformed how people consumed entertainment.

Profit is encouraged. Companies that excel in a sector will profit as their share of the market expands. Some of those profits benefit individuals or investors, while other capital is channeled back into the business to seed future growth. As markets expand, producers, consumers, and workers all benefit.

Bigger is often better. In economies of scale, large companies with easy access to large pools of capital and labor often enjoy an advantage over small producers that don't have the resources to compete. This condition can result in a producer driving rivals out of business by undercutting them on price or by controlling the supply of scarce resources, resulting in a market monopoly.

There are no guarantees. Unless a government chooses to intervene through market regulations or social welfare programs, its citizens have no promise of financial success in a market economy. Such pure laissez-faire economics is uncommon, though the degree of political and public support for such governmental intervention varies from nation to nation.

(Adapted from *What Is International Trade*?
https://wenku.baidu.com/view/ab3664872cc58bd63186bda4.html [2019-6-20])

☞New Words

1. border [ˈbɔːdə(r)] *n.* the boundary line or the area immediately inside the boundary 边境；边界；国界
2. political [pəˈlɪtɪkl] *adj.* involving or characteristic of politics or parties or politicians 政治的；党派的
3. industrialization [ɪnˌdʌstriəlaɪˈzeɪʃn] *n.* the development of industry on an extensive scale 工业化
4. transportation [ˌtrænspɔːˈteɪʃn] *n.* a facility consisting of the means and equipment necessary for the movement of passengers or goods 运输；运输系统；运输工具
5. globalization [ˌgləʊbəlaɪˈzeɪʃn] *n.* the process by which businesses or other organizations develop international influence or start operating on an international scale 全球化，全世界化
6. multinational [ˌmʌltiˈnæʃnəl] *adj.* involving or operating in many countries 跨国的；涉及多国的
7. outsource [ˈaʊtsɔːs] *v.* obtain goods or services by contract from an outside supplier; contract (work) out 外包；外购；外部采办
8. prevalence [ˈprevələns] *n.* the fact or condition of being prevalent; commonness 流行；普遍；广泛

9. Mercantilism [mɜːˈkæntɪlɪzəm] *n.* belief in the benefits of profitable trading; commercialism 重商主义；商业主义
10. paramount [ˈpærəmaʊnt] *adj.* more important than anything else; having the highest position or the greatest power 最重要的，首要的；至高无上的
11. multilateral [ˌmʌltiˈlætərəl] *adj.* agreed upon or participated in by three or more parties, especially the governments of different countries 多边的；多国的，多国参加的
12. socialist [ˈsəʊʃəlɪst] *adj.* adhering to or based on the principles of socialism 社会主义的
13. agricultural [ˌægrɪˈkʌltʃərəl] *adj.* relating to agriculture 农业的；农艺的
14. chiefly [ˈtʃiːfli] *adv.* for the most part 主要地；首先
15. protectionist [prəˈtekʃənɪst] *adj.* relating to the theory or practice of shielding a country's domestic industries from foreign competition by taxing imports
16. recession [rɪˈseʃn] *n.* a period of temporary economic decline during which trade and industrial activity are reduced 经济衰退；经济萎缩
17. domestic [dəˈmestɪk] *adj.* concerning the internal affairs of a country 国内的
18. regulation [ˌregjuˈleɪʃn] *n.* a principle or condition that customarily governs behavior 管理；规则
19. continent [ˈkɒntɪnənt] *n.* one of the large landmasses of the earth 大陆，洲，陆地

☞Phrases and Expressions

1. attempt to: to try to do sth. 尝试，企图；试图做某事
2. lack of: to be short of 没有，缺乏；不足
3. in some cases: under some circumstances 在某些情况下
4. engage in: to employ oneself in 从事/参加
5. rely on: to be dependent on, as for support or maintenance 依靠/依赖
6. in favor of: in support of 支持

☞Notes

1. Gross Domestic Product (GDP): a country's GDP is the total value of goods and services produced within a country in a year, not including its income from investments in other countries 国内生产总值
2. General Agreement on Tariffs and Trade (GATT): a multilateral international treaty signed in 1947 to promote trade, esp. by means of the reduction and elimination of tariffs and import quotas; replaced in 1995 by the World Trade Organization 关贸总协定
3. World Trade Organization: an international body concerned with promoting and regulating trade between its member states; established in 1995 as a successor to GATT 世界贸易

组织

4. North American Free Trade Agreement (NAFTA): an agreement for free trade between the United States and Canada and Mexico; became effective in 1994 for ten years 北美自由贸易协定

☞Exercises

Vocabulary

Replace the underlined words or expressions in the following sentences with one of the best choices from the word bank. Change the form when necessary.

cope	bother	chance	condemn	dress	dependent	come	reach	date
plunge	generate	leave	side	style	applicable			

1. She sometimes finds it difficult to <u>deal successfully with</u> all the pressure at work.
2. All parents <u>put the best clothes on</u> their children at the Spring Festival.
3. As soon as he finished composition, he handed it in; he didn't even <u>take trouble</u> checking its grammar and spelling.
4. After graduating from drama school, she <u>immediately began</u> her life as an actress.
5. The present controversy, which <u>began in</u> 1992, has lasted for more than ten years.
6. He <u>stretched out his hand to get</u> the phone and quickly dialed a number.
7. He was <u>given the punishment of</u> life imprisonment.
8. The letter asks him to consider the needs of older people who are <u>living on</u> state benefits.
9. What <u>matters</u> is that there are bad people out there, and somebody has to deal with them.
10. <u>It is most likely</u> that the patient will die within three months if she is not given proper treatment.

Unit 2　Humanities and Arts

✧ Text A

The Story of Steve Jobs

This is the text of the Commencement Address by Steve Jobs, CEO of Apple Computer and of Pixar Animation Studios, at Stanford University, delivered on June 12, 2005.

I am honored to be with you today at your commencement from one of the finest universities in the world. I never graduated from college. Truth be told, this is the closest I've ever gotten to college graduation. I dropped out of Reed College after the first 6 months, but then stayed around as a drop in for another 18 months or so before I really quit. So why did I drop out?

It started before I was born. My biological mother was a young, unwed college graduate student, and she decided to put me up for adoption. She felt very strongly that I should be adopted by college graduates, so everything was all set for me to be adopted at birth by a lawyer and his wife except that when I popped out they decided at the last minute that they really wanted a girl. So my parents, who were on a waiting list, got a call in the middle of the night asking, "We have an unexpected baby boy; do you want him?" They said, "Of course." My biological mother later found out that my mother had never graduated from college and that my father had never graduated from high school. She refused to sign the final adoption papers. She only relented a few months later when my parents promised that I would someday go to college. This was the start of my life.

And 17 years later I did go to college. But I naively chose a college that was almost as expensive as Stanford, and all of my working-class parents' savings were being spent on my college tuition. After six months, I couldn't see the value in it. I had no idea what I wanted to do with my life and no idea how college was going to help me figure it out. And here I was spending all of the money my parents had saved their entire life. So I decided to drop out and trusted that it would all work out OK. It was pretty scary at the time, but looking back it was one of the best decisions I ever made. The minute I dropped out I could stop taking the required classes that didn't interest me, and begin dropping in on the ones that looked far more interesting. It wasn't all romantic. I didn't have a dorm room, so I slept on the floor in friends' rooms. I returned coke bottles for the five-cent deposits to buy food with, and I would walk the 7 miles across town every Sunday night to get one good meal a week at the Hare Krishna temple. I loved it. And much of what I stumbled into by following my curiosity and intuition turned out to be priceless later on. Let me give you one example: Reed College at that time offered perhaps the best calligraphy instruction in the country. Throughout the campus, every poster, every label on every drawer, was beautifully hand calligraphed.

Because I had dropped out and didn't have to take the normal classes, I decided to take a calligraphy class to learn how to do this. I learned about serif and sanserif typefaces, about varying the amount of space between different letter combinations, about what makes great typography great. It was beautiful, historical, artistically subtle in a way that science can't capture, and I found it fascinating.

None of this had even a hope of any practical application in my life. But ten years later, when we were designing the first Macintosh computer, it all came back to me. And we designed it all into the Mac. It was the first computer with beautiful typography. If I had never dropped in on that single course in college, the Mac would have never had multiple typefaces or proportionally spaced fonts. And since Windows just copied the Mac, it's likely that no personal computer would have them. If I had never dropped out, I would have never dropped in on this calligraphy class, and personal computers might not have the wonderful typography that they do. Of course, it was impossible to connect the dots looking forward when I was in college. But it was very, very clear looking backwards ten years later.

Again, you can't connect the dots looking forward; you can only connect them looking backwards. So you have to trust that the dots will somehow connect in your future. You have to trust in something—your gut, destiny, life, karma, whatever. This approach has never let me down, and it has made all the difference in my life.

(Adapted from '*You've got to find what you love,' Jobs says.*
http://news.stanford.edu/news/2005/june15/jobs-061505.html[2018-10-8])

☞New Words

1. commencement [kəˈmensmənt] *n.* a ceremony at which students receive their academic degrees or diplomas 学位授予典礼，毕业典礼
2. animation [ˌænɪˈmeɪʃn] *n.* a film or movie in which drawings of people and animals seem to move 动画片
3. quit [kwɪt] 1) *vi/vt.* to leave your job, school, etc. 离开；离校 2) *vi/vt.* (informal) to stop doing sth. 停止，戒掉
4. biological [ˌbaɪəˈlɒdʒɪkl] 1) *adj.* connected with the processes that take place within living things 生物的；与生命过程有关的 2) *adj.* connected with the science of biology 生物学的
5. unwed [ʌnˈwed] *adj.* not married 没有结婚的，未婚的

6. adoption [əˈdɒpʃn] 1) *n.* the act of adopting a child 收养，领养 2) *n.* the decision to start using sth. such as an idea, a plan or a name 采纳，采用
7. relent [rɪˈlent] 1) *vi.* to finally agree to sth. after refusing 终于答应，不再拒绝 2) *vi.* to become less determined, strong, etc. 减弱，变缓和
8. scary [ˈskeəri] *adj.* causing fear or alarm 引起恐慌的
9. stumble [ˈstʌmbl] 1) *vi.* to move or walk in an unsteady way 跌跌撞撞地走，蹒跚而行 2) *vi.* to hit one's foot against sth. while one is walking or running and almost falls 绊脚
10. intuition [ˌɪntjuˈɪʃn] *n.* the ability to know sth. by using your feelings rather than considering the facts 直觉
11. calligraphy [kəˈlɪgrəfi] *n.* (the art of producing) beautiful handwriting 书法；书法艺术
12. serif [ˈserɪf] *n.* a short line at the top or bottom of some styles of printed letters 衬线，截线
13. typeface [ˈtaɪpfeɪs] *n.* a specific size and style of type within a type family（印刷用的）字体
14. typography [taɪˈpɒgrəfi] *n.* printing process 印刷术，排版，版面设计
15. historical [hɪˈstɒrɪkl] 1) *adj.* connected with the past（有关）历史的 2) *adj.* based on the study of history 有关历史研究的，历史学的
16. artistically [ɑːˈtɪstɪkli] *adv.* in an artistic manner 艺术性地；在艺术上
17. fascinating [ˈfæsɪneɪtɪŋ] *adj.* extremely interesting and attractive 极有吸引力的，迷人的
18. proportionally [prəˈpɔːʃənli] *adv.* corresponding in size or amount 成比例地，相应地
19. font [fɒnt] *n.* a specific size and style of type within a type family 字型，字体；铅字
20. destiny [ˈdestəni] 1) *n.* the power believed to control events 命运 2) *n.* what happens to sb./sth. (thought to be decided beforehand by fate) 天数，天命
21. karma [ˈkɑːmə] *n.* (Hinduism and Buddhism) the effects of a person's actions that determine his destiny in his next incarnation 因缘，因果报应

☞Phrases and Expressions

1. be honored (to do sth.): to feel proud and happy（做某事）感到荣幸
2. put (sb./sth.) up for: to recommend for 提供；推荐，提名
3. work out: 1) to develop in a successful way 成功地发展 2) to train the body by physical exercise 锻炼身体
4. stumble into sth.: to become involved in sth. by chance 无意间涉足某事
5. make all the difference: to have an important effect on sb./sth.有很大影响，使大不相同

☞Notes

1. Steve Jobs: Steve Jobs is an American entrepreneur and business magnate. He served as the chairman and chief executive officer (CEO) of Apple Inc., chairman of Pixar. He was

ever a member of The Walt Disney Company's board of directors following its acquisition of Pixar, and the chairman and CEO of NeXT. He is the co-founder of Apple Inc. and founder of NeXT. 史蒂夫·乔布斯是美国的企业家和商业巨头。他曾担任苹果公司的董事长兼首席执行官，皮克斯的董事长。收购皮克斯公司后，他曾是华特迪士尼公司的董事会成员，NeXT 的董事长兼首席执行官。他是苹果公司的联合创始人和 NeXT 公司的创始人。

2. Stanford University: Stanford University is a private research university in Stanford, California. Stanford is known for its academic strength, wealth, proximity to Silicon Valley, and ranking as one of the world's top universities. 斯坦福大学是加利福尼亚州的一所私立研究型大学。斯坦福以其学术实力、财力、靠近硅谷且身为世界顶尖大学而闻名。

3. Pixar Animation Studios: Pixar Animation Studios, a wholly-owned subsidiary of The Walt Disney Company, is an Academy Award-winning film studio with world-renowned technical, creative and production capabilities in the art of computer animation and creators of some of the most successful and beloved animated films of all time. 皮克斯动画工作室是华特迪士尼公司的全资子公司，是一家获得奥斯卡奖的电影工作室。该工作室在计算机动画艺术领域拥有世界知名的技术、创意和制作能力，也创作了一些大获成功并深受观众喜爱的动画影片。

4. Reed College: Reed College is an independent liberal arts college in southeast Portland in the U.S. state of Oregon. Founded in 1908, Reed is a residential college with a campus in Portland's Eastmoreland neighborhood, featuring architecture based on the Tudor-Gothic style, and a forested canyon nature preserve at its center. 里德学院是一个位于美国俄勒冈州东南波特兰的独立的文理学院。它成立于 1908 年，是一所住宿学院，在波特兰的 Eastmoreland 社区设有校园，其建筑基于都铎-哥特式风格，在其中心有一个森林覆盖的峡谷自然保护区。

5. Hare Krishna: Hare Krishna is the popular name for the International Society of Krishna Consciousness (or ISKCON). Hare Krishna 是国际奎师那知觉协会的常用名称。

6. Macintosh: The Macintosh (pronounced as /ˈmækɪnˌtɒʃ/ MAK-in-tosh; branded as Mac since 1998) is a family of personal computers designed, manufactured, and sold by Apple Inc. Macintosh（发音为麦金托什；自 1998 年以来被称为 Mac）是由苹果公司设计、制造和销售的一款个人计算机。

☞Exercises

Ⅰ. Reading Comprehension

Answer the following questions according to the text.

1. What did Steve Jobs do after he dropped out of Reed College?

2. What can we learn from Steve Jobs' narration about his biological mother?
3. What kind of people were Steve Jobs' adoptive parents?
4. Why did Steve Jobs drop out of Reed College?
5. What did Steve Jobs think about his decision of dropping out? Why?

Ⅱ. Vocabulary

A. Fill in the blanks in the paragraph by selecting the most suitable words from the word bank.

A. enough	B. fit	C. emphasis	D. practical
E. innumerable	F. concentrate	G. adopt	H. questionable
I. profound	J. factor	K. too	L. substance
M. passion	N. emotion	O. fix	

Personality is, to a large extent, inherent—A-type parents, usually bring about A-type children. But the environment must also have a __1__ effect, where if competition is important to the parents it is likely to become a major __2__ in the lives of their children.

One place where children soak up A characteristics is school, which is, by its very nature, a highly competitive institution. Too many schools __3__ the "win at all costs" moral standard and measure their success by sporting achievements. The current __4__ for making children compete against their classmates or against the clock produces a two-layer system, in which competitive A-types seem in some way better than their B-type fellows. Being __5__ keen to win can have dangerous consequences: remember that Pheidippides, the first marathon runner, dropped dead seconds after saying; "cheers, we conquer!"

By far the worst form of competition in schools is the extreme __6__ on examinations. It is a rare school that allows pupils to __7__ on those things they do well. The merits of competition by examination are somewhat __8__, but competition in the certain knowledge of failure is positively harmful.

Obviously, it is neither __9__ nor desirable that all A youngsters change into B's. The world needs types, and schools have an important duty to try to __10__ a child's personality to his possible future employment. It is top management.

B. Fill in the blanks in each sentence by selecting the most suitable words.

1. Betty advised me to label our luggage carefully in case it gets ____ in transit.
   A. misused   B. mishandled   C. mistaken   D. mislaid
2. This house will probably come on the ____ next month.
   A. fair   B. market   C. shop   D. store
3. An institution that properly carries the name university is a more comprehensive and

complex institution than any other kind of higher education ____.

A. settlement B. establishment C. construction D. structure

4. Jack is so ____ to his appearance that he never has his clothes pressed.

A. adverse B. anonymous C. indifferent D. casual

5. Outside my office window there is a fire ____ on the right.

A. escape B. ladder C. steps D. stairs

6. The electric fan does not work because of the ____ of service.

A. pause B. break C. interruption D. breakdown

7. The thieves ____ the waste paper all over the room while they were searching for the diamond ring.

A. spread B. scratched C. scattered D. burned

8. The ____ problem of bringing a space ship back from the moon has been solved.

A. technical B. technological C. technique D. technology

9. I was awfully tired when I got home from work, but a half-hour nap ____ me.

A. revived B. released C. relieved D. recovered

10. Information and opinion gap exercises have to have some content ____ talking about.

A. worthwhile B. worthily C. worth D. worthy

Ⅲ. Cloze

Choose an appropriate word from the four choices marked A, B, C and D for each blank in the passage.

Practically all people __1__ a desire to predict their future circumstances. People seem inclined to __2__ this task using causal reasoning. First, we generally recognize that future circumstances are somehow caused or conditioned by present __3__. We learn that getting an education will affect how much money we earn later in life and that swimming beyond the reef may bring an unhappy __4__ with a shark.

Second, people also learn that such __5__ of cause and effect are probabilistic in nature. That is, the effects occur more often when the causes occur than when the causes are absent—but not always.

Thus, students learn that studying hard __6__ good grades in most instances, but not every time. Science makes these concepts of causality and probability more explicit and provides techniques for dealing __7__ them more rigorously than does causal human inquiry. It sharpens the skills we already have by making us more conscious, rigorous, and explicit in our inquiries.

In looking at the ordinary human inquiry, we need to __8__ between prediction and understanding. Often, we can make predictions without understanding. And often, even if we don't understand why, we are willing to act on the basis of a demonstrated predictive ability.

Whatever the primitive drives or instincts that motivate human beings, satisfying them depends heavily on the ability to predict future circumstances. The attempt to predict is often played in a context of knowledge and understanding. If you can understand why things are related to one another, why certain regular patterns __9__, you can predict better than if you simply observe and remember those patterns. Thus, human inquiry aims __10__ answering both "what" and "why" questions, and we pursue these goals by observing and figuring out.

1. A. exhibit	B. exaggerate	C. examine	D. exceed
2. A. underestimate	B. undermine	C. undertake	D. undergo
3. A. one	B. ones	C. one's	D. oneself
4. A. meeting	B. occurrence	C. encounter	D. sighting
5. A. patterns	B. designs	C. arrangements	D. pictures
6. A. producing	B. produces	C. produce	D. produced
7. A. for	B. at	C. in	D. with
8. A. distinguish	B. distinct	C. distort	D. distract
9. A. happen	B. occur	C. occupy	D. incur
10. A. at	B. on	C. to	D. beyond

Ⅳ. Translation

Translate the following text into Chinese.

Acupuncture is an important part of traditional Chinese medicine (TCM). In accordance with the "main and collateral channels" theory in TCM, the purpose of acupuncture is to dredge the channel and regulate qi and blood, so as to keep the body's yin and yang balanced and achieve reconciliation between the internal organs. It features in traditional Chinese medicine that "internal diseases are to be treated with external therapy". The main therapy of acupuncture involves using needles to pierce certain acupoints of the patient's body, or adopting moxibustion to stimulate the patient's acupoints so as to stimulate the channels and relieve pain. With its unique advantages, acupuncture has been handed down generation after generation and has now spread all over the world. Nowadays, acupuncture, along with Chinese food, kung fu (otherwise known as Chinese martial arts), and traditional Chinese medicine, has been internationally hailed as one of the "four new national treasures".

Ⅴ. Writing

Write an essay commenting on Albert Einstein's remark "I have no special talents, I am only passionately curious." You can cite one example or two to illustrate your point of view. You should write at least 150 words.

✧ Text B

Utopia of Reason

A moral psychologist argues for setting aside feelings in favor of facts

In an age of partisan divides, it has become popular to assert that the wounds of the world would heal if only people made the effort to empathize more with each other. If only white police officers imagined how it feels to be a black man in America; if only black Americans understood the fears of the man in uniform; if only Europeans opposed to immigration walked a mile in the shoes of a Syrian refugee; if only tree-hugging liberals knew the suffering of the working class.

Barack Obama warned of an empathy "deficit" in 2006, and did so again in his valedictory speech in January: "If our democracy is to work in this increasingly diverse nation," he said, "each one of us must try to heed the advice of one of the great characters in American fiction, Atticus Finch, who said, 'You never really understand a person until you consider things from his point of view...until you climb into his skin and walk around in it.'"

It is a piece of generous, high-minded wisdom with which few would dare to disagree. But Paul Bloom, a psychologist at Yale University, does disagree. His new book, *Against Empathy*, makes the provocative argument that the world does not need more empathy; it needs less of it. People are bingeing on a sentiment that does not, on balance, make the world a better place. Empathy is "sugary soda, tempting and delicious and bad for us". In its stead, Mr. Bloom prescribes a nutritious diet of reason, compassion and self-control.

To be clear, Mr. Bloom is not against kindness, love or general goodwill toward others. Nor does he have a problem with compassion, or with "cognitive" empathy—the ability to understand what someone else is feeling. His complaint is with empathy defined as feeling what someone else feels. Though philosophers at least as far back as Adam Smith have held it up as a virtue, Mr. Bloom says it is a dubious moral guide. Empathy is biased: people tend to feel for those who look like themselves. It is limited in scope, often focusing attention on the one at the expense of the many, or on short-term rather than long-term consequences. It can incite hatred and violence—as when Donald Trump used the example of Kate Steinle, a woman murdered by an undocumented immigrant, to drum up anti-immigrant sentiment, or when Islamic State fighters point to instances of Islamophobia to encourage terrorist attacks. It is innumerate, blind to statistics and to the costs of saccharine indulgence.

Empathy can be strategically useful to get people to do the right thing, Mr. Bloom acknowledges, and it is central to relationships (though even here it must sometimes be overridden, as any parent who takes a toddler for vaccinations knows). But when it comes to policy, empathy is too slippery a tool. "It is because of empathy that citizens of a country can be transfixed by a girl stuck in a well and largely indifferent to climate change," he writes. Better to rely on reason and cost-benefit analysis. As rational arguments for environmental protection or civil rights show, morality is possible without sentimental appeals to individual suffering. "We should aspire to a world in which a politician appealing to someone's empathy would be seen in the same way as one appealing to people's racist bias," Mr. Bloom writes. Racism, like anger or empathy, is a gut feeling; it might be motivating, but that kind of thinking ultimately does more harm than good.

That is a radical vision—and like many Utopias, one with potentially dystopian consequences. Unless humans evolve into something like the Vulcans from "Star Trek", guided purely by logic, it is also unimaginable. Reason should inform governance, but people tend to be converted to a cause—gay marriage, for instance—by emotion. Yet Mr. Bloom's point is a good one: empathy is easily exploited, marshalled on either side of the aisle to create not a bridge but an impasse of feelings. In a time of post-truth politics, his book offers a much-needed call for facts.

(Adapted from *The Economist*, 2018(2), 73-74. by Peter Bloom)

☞New Words

1. dystopia [dɪsˈtəʊpiə] *n.* an imaginary place where life is extremely difficult and a lot of unfair or immoral things happen 反面乌托邦，反面假想国
2. partisan [ˌpɑːtɪˈzæn] *n.* a person who strongly supports a particular leader, group or idea 坚定的支持者，铁杆拥护者
3. assert [əˈsɜːt] *v.* to state firmly that something is true（坚决）主张；断言
4. liberal [ˈlɪbərəl] 1) *n.* a person who supports political, social and religious change 支持（社会、政治或宗教）变革的人 2) *n.* a person who understands and respects other people's opinions and behaviour, especially when they are different from their own 理解且尊重他人意见的人；宽容的人；开明的人 3) *n.* Liberal (politics), a member of the British Liberal Party in the past, or of a Liberal Party in another country（政治）（旧时）英国自由党成员；（英国以外国家的）自由党成员
5. heed [hiːd] 1) *v.* to pay attention to someone's advice or warning 注意（别人的建议或警告）2) *n.* careful attention 留心；注意；听从
6. high-minded [ˌhaɪˈmaɪndɪd] *adj.* having very high moral standards or principles 品德（情操）高尚的，高洁的
7. provocative [prəˈvɒkətɪv] 1) *adj.* causing anger or another strong reaction 使人生气的；挑衅的；煽动的；引起争论的 2) *adj.* intended or intending to make someone sexually excited 引诱的；激起性欲的
8. binge [bɪndʒ] *n.* (informal) a short period when you do too much of something, such as

eating or drinking 狂欢作乐；大吃大喝

9. prescribe [prɪˈskraɪb] *v.* to say what medicine or treatment a sick person should have 给……开药（开处方）
10. goodwill [ˌgʊdˈwɪl] *n.* friendly or helpful feelings towards other people or countries 友善；友好；善意；亲善
11. incite [ɪnˈsaɪt] *v.* to deliberately encourage people to fight, argue etc. 煽动，鼓动，激起
12. innumerate [ɪˈnjuːmərət] *adj.* unable to do simple calculations or understand basic mathematics 不会计算的；不会基础数学的
13. saccharine [ˈsækəriːn] *adj.* (disapproving)(of people or things) too emotional in a way that seems exaggerated（人或物）情感过分强烈而显夸张的；故作多情的（同 sentimental）
14. indulgence [ɪnˈdʌldʒəns] *n.* (usually disapproving) the state or act of having or doing whatever you want; the state of allowing sb. to have or do whatever they want 沉溺；放纵；纵容
15. override [ˌəʊvəˈraɪd] *v.* to be regarded as more important than something else 视为比……更重要
16. toddler [ˈtɒdlə(r)] *n.* a child who has only recently learnt to walk 学步的儿童;刚学会走路的孩子
17. rational [ˈræʃnəl] *adj.* based on or in accordance with reason or logic 合理的，基于理性的
18. civil rights [ˌsɪvlˈraɪts] *n.*[pl] the rights that people have in society to equal treatment and equal opportunities, whatever their race, sex, or religion 公民权
19. aspire [əˈspaɪə(r)] *v.* to desire and work towards achieving something important 追求，渴望，有志于
20. motivating [ˌməʊtɪˈveɪtɪŋ] *adj.* providing an incentive or impelling to action 激励人心的；激励的；鼓舞人心的
21. evolve [iˈvɒlv] *v.* to develop and change gradually over a long period of time 逐步发展；逐渐演变
22. convert [kənˈvɜːt] *v.* to change or make sb. change their religion or beliefs（使）改变（宗教或信仰）；（使）皈依，归附
23. marshall [ˈmɑːʃl] 1) *v.* (also marshal) to organize your thoughts, ideas etc. so that they are clear, effective, or easy to understand 整理（思路、想法等）2) *v.* to organize all the people or things that you need in order to be ready for a battle, election etc. 调集，集结（力量） 3) *v.* to control or organize a large group 控制；组织；引领
24. impasse [ˈɪmpɑːs] *n.*[usually sing.] a difficult situation in which no progress can be made because the people involved cannot agree on what to do 僵局；绝境（同 deadlock、stalemate）

☞Phrases and Expressions

1. gut reaction/feeling/instinct (informal): a reaction or feeling that you are sure is right, although you cannot give a reason for it 本能的反应/本能的感觉/直觉
2. argue for/against (doing) sth.: to give support to/ object to sth. 支持/反对某事
3. set sth. aside: to decide not to consider a particular feeling or thing because something else is more important 暂不考虑，不顾；对……置之不理
4. in favor (of sb./sth.): if you are in favor of sb./sth., you support and agree with them/it 赞同；支持
5. in uniform: 1) wearing a uniform 穿着制服 2) in the Army, Navy etc. 做一名军人，当兵
6. in sb.'s shoes: in someone else's situation, especially a bad one 处于某人的境地（尤指恶劣处境）
7. tree-hugger: someone who wants to protect the environment, and who takes part in protests which some people think are silly (used showing disapproval)（如抱住树木以使其免遭砍伐等的）环境保护者，环保主义分子，环保论者（含贬义）
8. general goodwill toward others: be kind 与人为善
9. feel for sb.: to feel sympathy for someone 同情（某人）
10. drum sth. up: to try hard to get support or business 竭力争取（支持）；兜揽（生意）
11. be indifferent: to have a complete lack of interest in sb./sth. 对……不关心的，不在乎的
12. cost-benefit analysis (CBA): Cost-benefit analysis, sometimes called benefit-cost analysis (BCA), is a systematic approach to estimating the strengths and weaknesses of alternatives that satisfy transactions, activities or functional requirements for a business. 成本效益分析；本利分析

☞Notes

1. Utopia: an imaginary place or state in which everything is perfect 乌托邦；空想的完美境界（源自托马斯·莫尔爵士所著的书名，书中描绘了这样一个地方。）
2. Atticus Finch: Atticus Finch is a fictional character in author Harper Lee's Pulitzer Prize winning novel of 1960, *To Kill a Mockingbird*. 阿迪克斯·芬奇是作家哈珀·李（Harper Lee）1960 年普利策奖获奖小说《杀死一只知更鸟》中的虚构角色。
3. Vulcan: Vulcans (also Vulcanians) are a fictional extraterrestrial humanoid species in the Star Trek franchise who originate from the planet Vulcan. In the various Star Trek television series and movies, they are noted for their attempt to live by logic and reason with as little interference from the emotion as possible. They were the first extraterrestrial species in the Star Trek universe to observe the first contact protocol with humans. 瓦肯人是虚构科幻电视剧《星际迷航》中的一种外星人。他们是发源于瓦肯星的智慧外

星类人类族群。在各种《星际迷航》电视连续剧和电影中，他们以信仰严谨的逻辑和推理，不受任何情感的干扰闻名。它们是星际迷航宇宙中观察到的第一个与人类接触的外星物种。

4. post-truth politics: Post-truth politics (also called post-factual politics and post-reality politics) is a political culture in which debate is framed largely by appeals to emotion disconnected from the details of policy, and by the repeated assertion of talking points to which factual rebuttals are ignored. 后真相政治（又称后事实政治或后现实政治）是一种政治文化，在这种政治文化中,人们很大程度上关注与政策细节脱节的情感的呼吁，以及反复断言的观点，而对事实的反驳则被忽视。后真相政治是“事实胜于雄辩”的相反面，即是“雄辩胜于事实”，意见重于事实，立场决定是非；人们把情感和感觉放在首位，证据、事实和真相沦为次要（甚至毫不重要）；政治人物说谎，不再是为了瞒骗，而是巩固目标群众的偏见，换取共鸣与支持。

☞Exercises

Vocabulary

Fill in the blanks in the paragraph by selecting the most suitable words from the word bank.

A. cancel	B. pace	C. extreme	D. automatically
E. remove	F. vital	G. performance	H. supposed
I. rate	J. exposure	K. achievement	L. unusual
M. obviously	N. withstand	O. harsh	

As the __1__ of life continues to increase, we are fast losing the art of relaxation. Once you are in habit of rushing through life, being on the go from morning till night, it is hard to slow down. But relaxation is essential for a healthy mind and body.

Stress is a natural part of everyday life and there is no way to avoid it. In fact, it is not the bad thing it is often __2__ to be. A certain amount of stress is __3__ to provide motivation and give purpose to life. It is only when the stress gets out of control that it can lead to poor __4__ and ill health.

The amount of stress a person can __5__ depends very much on the individual. Some people are not afraid of stress, and such characters are __6__ prime material for managerial responsibilities. Others lose heart at the first signs of __7__ difficulties. When exposed to stress, in whatever form, we react both chemically and physically. In fact, we make choices between “fight” or “flight” and in more primitive days the choices made the difference between life or death. The crises we meet today are unlikely to be so __8__, but however little the stress, it involves the same response. It is when such a reaction lasts long, through continued __9__ to stress, that health becomes endangered. Such serious conditions as high blood pressure and heart disease have established links with stress. Since we cannot __10__ stress from our lives (it would be unwise to do so even if we could), we need to find ways to deal with it.

Unit 3 Information and Technology

✧ Text A

WeChat Launched Its Digital Revolution

Tencent's WeChat, China's most popular social media app, has launched its digital payments platform in Malaysia, its first market in Asia beyond China. Malaysia's WeChat users will be able to transfer money among themselves and make payments to offline merchants in ringgit (林吉特，马来西亚货币单位). This suggests Tencent is building a local payments service, rather than taking the more common route of overseas expansion used by Chinese mobile-app providers that cater to Chinese tourist or nationals living abroad.

Malaysia's central bank has been implementing policies promoting electronic payments in a bid to boost a network that lags behind other south-east Asian markets. That has triggered the launch of digital wallets by other strong players, including Grab, the south-east Asia ride-hailing company. "Malaysia is a vibrant market. Technology-savvy Malaysians are embracing a digital lifestyle and to meet this shift, the payment experience has to evolve. Bringing WeChat Pay to Malaysia is our response to this," said WeChat Pay Malaysia.

SY Lau, senior vice-president at Tencent, told Reuters in November when the company acquired a Malaysian epayment licence, that WeChat had 20 million users in the country, equivalent to almost two-thirds of the population. The potential for mobile payments is vast in Malaysia, where cash is still king but the number of mobile phones, mostly smartphones, outstrips a population of 32.1 million by more than 10 million, according to the central bank. But collaborations with local banks, of which WeChat has none, will be just as important for WeChat Pay to flourish there. At home, it took Tencent and Ant Financial, Alibaba's electronic payments affiliate, years to build the links with hundreds of Chinese banks that make their services possible.

Tencent Holding Ltd, a Shenzhen-based provider of Internet services, will speed up the commercialization process of its popular WeChat messaging application as part of its global expansion strategy, the company's top manager said. Ma Huateng, Tencent's CEO, said the company is planning to build an open platform so that Internet service developers are able to provide more features for WeChat users. "We are developing value-added services for the application, especially in the mobile social contacts and online games sectors," Ma, who is also a deputy from Shenzhen, said during the annual session of National People's Congress in Beijing. WeChat, which is similar to the United States' WhatsApp and to some other mobile apps, enables users to chat via text messages, voice and video on their mobile devices as long as they have a web connection. WeChat opened a group purchase service this month, and the application will add more value-added services in the near future, "It's a very complicated process to develop services for WeChat, given that it involves a lot of work in online and offline integration. We're determined to cooperate with other Internet service providers to build a business model," Ma said.

Industry insiders said that Tencent has vast potential to develop more services affiliated with the brand and can increase its global market presence, given that it registered fast growth in revenue in recent years. Sources with the company said that its sales revenue reached 11.6 billion yuan ($1.9 billion) in the third quarter of last year, up 54.3 percent year-on-year. With a growing number of WeChat users at home and abroad, Tencent is moving ahead with its overseas expansion plans. It has set up an office in the US, the first overseas unit of its kind, to boost the development of WeChat. "We're facing a big challenge in the overseas market, especially in the US. Many mobile apps like WhatsApp and Facebook Messenger have a strong local presence. Ma said, "We are studying US users' habits and cooperating with local developers to provide different experiences for overseas users." In a motion submitted to the National People's Congress, Ma said that Chinese Web companies are ready to expand overseas after nearly two decades of development.

"A growing number of Chinese companies in the manufacturing industry have opened businesses overseas during the past several years. China needs to further expand its overseas investments by encouraging more businesses engaged in the services industry, especially in the information and communication technologies, to invest in foreign markets," Ma told. The global expansion by Chinese Web companies will help increase the value of China's services trade in the international market. According to the Ministry of Commerce, the services industry currently accounts for nearly 70 percent of the global economy, but China still has a trade deficit in the services sector. "More investments in the overseas market from China's Internet companies will help reduce the country's services trade deficit," Ma expects.

(Adapted from *sinaEnglish* 2013.3.
http://english.sina.com/business/2013/0312/570509.html[2018-9-1])

☞New Words

1. affiliate [əˈfɪlieɪt] 1) *n.* a subsidiary or subordinate organization that is affiliated with

another organization 附属机构，分公司 2) *vt.* to officially attach or connect (a subsidiary group or person) to an organization 使隶属于，加入

2. expansion [ɪkˈspænʃn] *n.* the act of increasing (something) in size or volume or quantity or scope 扩大，膨胀，扩充
3. network [ˈnetwɜːk] (electronics) *n.* a system of interconnected electronic components or circuits, or a system of intersecting lines or channels 网络，网状物，网状系统
4. boost [buːst] 1) *vt.* to increase or contribute to the progress or growth of 推进，提高，增加 2) *n.* a source of help or encouragement leading to increase or improvement 推动，促进
5. response [rɪˈspɒns] *n.* a result or a statement (either spoken or written) that is made in reply to a question or request or criticism or accusation 回答，响应，反应，答复
6. transfer [trænsˈfɜːr] 1) *n.* the act of moving something from one location to another; a ticket that allows a passenger to change conveyances 迁移，移动，换车 2) *v.* to move someone or something from one place to another 转移，调转，调任
7. evolve [iˈvɒlv] *v.* to work out; to undergo development or evolution; to gain through experience 进展，进化，展开
8. platform [ˈplætfɔːm] *n.* a raised horizontal surface; a document stating the aims and principles of a political party 平台，站台，月台，讲台，（政党的）政纲
9. route [ruːt] *n.* an established line of travel or access; an open way (generally public) for travel or transportation 路线，（固定）线路，途径
10. vast [vɑːst] *adj.* unusually great in size or amount or degree or especially extent or scope 巨大的，广阔的

☞Phrases and Expressions

1. rather…than: instead of or would rather do something than to do something 而不是；宁愿……而不愿……
2. bring…to: to take something to some place or to restore (a person) to consciousness 带来；使（某人）恢复意识
3. equivalent to: equal to / up to 等于，相当于；与……等值

☞Notes

1. Grab: Grab is a Singapore-based technology company offering ride-hailing transport services, food delivery and payment solutions, and a very large taxi service platform in Southeast Asia. Grab 是一家总部位于新加坡的科技公司，提供乘驾运输服务、食品配送和支付服务，是东南亚非常大的打车服务平台。
2. Ant Financial: Ant Financial was founded in 2014 and started from Alipay (支付宝),

provides financial services for small enterprise and individual consumers based on mobile Internet, big data and cloud computing. 蚂蚁金服集团成立于2014年，起步于支付宝，依靠移动互联、大数据、云计算为基础，为小微企业和个人消费者提供金融服务。

3. Alibaba: Alibaba was found in 1998, and operates much business, including Taobao, Tmall, Alibaba International Trading Market, Ali Cloud, Ant Financial, 1688, etc. 阿里巴巴集团成立于1998年，经营多项业务，包括淘宝、天猫、阿里巴巴国际交易市场、阿里云、蚂蚁金服、1688等。

☞Exercises

Ⅰ. Reading Comprehension

1. What convenience could the payment platform bring to Malaysian WeChat users?
2. What does the attitude of the Malaysian take to electronic payment?
3. How about the measures?
4. How can Malaysia learn from the development of electronic payment in China?

Ⅱ. Vocabulary

A. Fill in the blanks in the paragraph by selecting the most suitable words from the word bank.

A. interested	B. frustrating	C. comfortable	D. valuable	E. post	F. virtual
G. rewarding	H. insights	I. embarrassed	J. communicate	K. benefit	L. medium
M. explanation	N. information	O. minimum			

Good communication is the key to success when learning online. You should take the opportunity to get to know your teacher and classmates through email and by participating in Internet discussions. This will lead to a more positive and __1__ learning experience.

It's true that learning the technology needed to take part in a class can at times be __2__. For example, you may need to ask how to __3__ your assignment on the Web. But don't worry! If you have a problem, ask for help. There's no such thing as a stupid question, so there is no need to be intimidated or __4__. Sharing __5__ and answer freely is what makes the Internet such a great medium for learning.

Online classroom teacher Mike Roberts was asked about what he thought the greatest __6__ of online learning is. "As a teacher, I need the students to ask questions so that I know what areas of my lessons need further __7__. That's what is great about teaching and learning over the Internet. In an ordinary classroom, time is limited, so students seem to ask the __8__ amount of questions possible. But in the __9__ classroom, students are always asking questions. They really seem to feel __10__ asking me for the information that they need. They also share a lot of valuable ideas with each other in a way that you don't usually see in a regular classroom."

B. Fill in the blanks in each sentence by selecting the most suitable words.

1. I ____ on the phone, that she was very upset, and her voice changed a little.
   A. recognized B. conceived C. deceived D. perceived
2. Every chemical change either results from energy being used to produce the change or causes energy to be ____ in some form.
   A. given off B. put out C. set off D. used up
3. We had to walk with ____ on that dangerous, icy slope.
   A. caution B. precaution C. ease D. suspicion
4. To ____ for his unpleasant experiences, he drank a little more than was good for him.
   A. commence B. compromise C. compensate D. compliment
5. He was such a ____ speaker that he held our attention every minute of the three-hour lecture.
   A. specific B. dynamic C. heroic D. diplomatic
6. The group of technicians are engaged in a study which ____ all aspects of urban planning.
   A. inserts B. grips C. performs D. embraces
7. The struggle of ____ within him was so tense that he could hardly speak.
   A. emotions B. motions C. moods D. enthusiasm
8. The ginger should be ____ before it is added to the boiling jam.
   A. shattered B. cracked C. smashed D. crushed
9. American companies are evolving from mass-manufacturing to ____ enterprises.
   A. moveable B. changing C. flexible D. varying
10. The government has devoted a larger slice of its national ____ to agriculture than most other countries.
    A. resources B. potential C. budget D. economy
11. After looking through the students' homework, he ____ to have the history class.
    A. proceeded B. processed C. presided D. preceded
12. The directions were so ____ that it was impossible to complete the assignment.
    A. ingenious B. ambitious C. notorious D. ambiguous
13. My boss often ____ his opinion upon others.
    A. imports B. imposes C. exposes D. disposes
14. The holidays are ____ over; there's only one day left.
    A. practical B. practically C. practicable D. practicably
15. I'm afraid this painting is not by Picasso. It's only a copy and so it's ____.
    A. priceless B. invaluable C. unworthy D. worthless

16. With prices ____ so much, it's hard for the company to plan a budget.
    A. fluctuating B. waving C. swinging D. vibrating

Ⅲ. Cloze

Choose an appropriate word from the four choices marked A, B, C and D for each blank in the passage.

If you were to begin a new job tomorrow, you would bring with you some basic strengths and weaknesses. Success or __1__ in your work would depend, to __2__ great extent, __3__ your ability to use your strengths and weaknesses to the best advantage.

__4__ the utmost importance is your attitude. A person __5__ begins a job convinced that he isn't going to like it or is __6__ that he is going to ail is exhibiting a weakness which can only hinder his success. On the other hand, a person who is secure __7__ his belief that he is probably as capable __8__ doing the work as anyone else and who is willing to make a cheerful attempt __9__ it possesses a certain strength of purpose. The chances are that he will do well. __10__ the prerequisite skills for a particular job is strength. Lacking those skills is obviously a weakness. A bookkeeper who can't add or a carpenter who can't cut a straight line with a saw __11__ hopeless cases. This book has been designed to help you capitalize __12__ the strength and overcome the __13__ that you bring to the job of learning. But in groups to measure your development, you must first __14__ stock of somewhere you stand now. __15__ we get further along in the book, we'll be __16__ in some details with specific processes for developing and strengthening __17__ skills. However, __18__ begin with, you should pause __19__ examine your present strengths and weaknesses in three areas that are critical to your success or failure in school: your __20__, your reading and communication skills, and your study habits.

1. A. improvement B. victory C. failure D. achievement
2. A. a B. the C. some D. certain
3. A. in B. on C. of D. to
4. A. Out of B. Of C. To D. Into
5. A. who B. what C. that D. which
6. A. ensure B. certain C. sure D. surely
7. A. onto B. on C. off D. in
8. A. to B. at C. of D. for
9. A. near B. on C. by D. at
10. A. Have B. Had C. Having D. Had been
11. A. being B. been C. are D. is
12. A. except B. but C. for D. on
13. A. idea B. weakness C. strength D. advantage

14. A. make	B. take	C. do	D. give
15. A. As	B. Ill	C. Over	D. Out
16. A. deal	B. dealt	C. be dealt	D. dealing
17. A. learnt	B. learned	C. learning	D. learn
18. A. around	B. to	C. from	D. beside
19. A. to	B. onto	C. into	D. with
20. A. intelligence	B. work	C. attitude	D. weakness

Ⅳ. Translation

Translate the following text into Chinese.

Over 90% of mobile apps gained permission to access users' private information in the first half of 2018, according to a report recently released by Tencent's Research Center on Society and the Data Center of China Internet (DCCI). Statistics indicate that about 99.9% of Android apps obtained private information from their users in the first six months of this year. The figure for iOS apps also rose dramatically from 69.3% in the first half of 2017 to 93.8%. Of them, game apps saw the biggest increase in gaining access to users' private data, from 43.1% in the second half of last year to 88.9% in the first half of 2018. Location information was required by 89.3% of apps in the first half of 2017, and the figure saw an increase to 95.9% in the first six months of this year, said Hu Yanping, founder of DCCI, adding that apps requiring access to users' contact data also climbed from 43.7% to 61.2% during the same period. Although relevant laws and regulations have been introduced in China, many apps are still suspected of overusing their authorization to collect user data, said Zhou Ye, senior mobile security researcher at 360 Vulpecker Team. Huang Xiaolin, a director of Tencent, also stressed that internet companies should self-discipline when it comes to the collection of information, usage of information and notification to the users.

Ⅴ. Writing

Directions: For this part, you are allowed to write a short essay on living in the virtual world. Try to imagine what will happen when people spend more and more time in the virtual world instead of interacting in the real world. You are required to write at least 150 words but no more than 200 words.

✧ Text B

Apple: From iPhones to iCars

Apple is entering the auto business, but the road ahead could be rough. Having redefined the personal-computer and mobile-phone industries, Apple has set its sights on a new, moving target. The darling of the tech industry is aiming to debut an electric car in 2019, according to a report by the Wall Street Journal.

For years there had been speculation that Apple had auto ambitions. It has been hiring swarms of engineers to work on the project, code-named Titan, which now employs around 600. With around $200 billion stored up, Apple certainly has enough cash to spend on a new venture.

Its legions of fans are always eager to see it launch a new offering that makes a splash. Its most recent product, the Apple Watch, has not been the massive hit some analysts expected, and Apple's launch event earlier this month was rather dull, showcasing mainly tweaks to existing products. Launching a car would be far more daring and, in theory, highly lucrative. Global car sales were worth around $2 trillion last year. However, there are many things about the car business that make it quite unlike peddling phones. The replacement cycle for an iPhone is a mere two years. Consumers hold on to their cars for far longer. To make business sense, Apple's cars would need to earn returns comparable to those of its existing products. But gross profit margins like Apple's current 40% or so are something most large-scale carmakers can only dream of. At BMW, one of the most profitable of these, such margins are around 20%. An even more important factor is that, as demonstrated by all this week's attention on Volkswagen, cars are more heavily regulated than consumer electronics. Compliance with safety standards and emissions rules is likely to be more rigorously enforced in the future. That is a tricky prospect for even the most experienced carmaker, let alone an industry newbie.

The firm's capabilities in mobile devices will be an advantage, as cars become ever more connected to the internet. But the risks associated with connected cars are of a different order to those with mobile phones. This week it was revealed that malware had infiltrated some of the smartphone apps sold in Apple's store, including two of the most popular apps in China. This was embarrassing but not disastrous. If similar malware got into an Apple car, the results could be deadly. Such concerns are not enough to discourage tech firms from driving at full speed into the car business.

Leading the way is Tesla, a maker of upmarket electric vehicles set up by Elon Musk, a successful tech entrepreneur. Google is working on a self-driving vehicle and an operating system for cars, and recently hired a former motor-industry executive to run its autonomous car project. Apple's decision to enter the motor industry may be fuelled in part by not wanting to leave Google with control of the dashboard and cars' operating systems. Their rivalry is turning into a road race.

(Adapted from *The Economist*. 2017(1). by Smith Clover,
http://www.kekenet.com/menu/201702/493059.shtml [2017-2-25])

☞New Words

1. vehicle [ˈviːəkl] *n.* a conveyance that transports people or objects 车辆，交通工具
2. decision [dɪˈsɪʒn] *n.* a position or opinion or judgment reached after consideration 决定，决策
3. disastrous [dɪˈzɑːstrəs] *adj.* causing great damage 灾难性的
4. debut [ˈdeɪbjuː] 1) *n.* first public performance, appearance, or recording 初次登场，首次露面 2) *v.* to appear in public or become available for the first time 初次登场
5. unusual [ʌnˈjuːʒuəl] *adj.* not habitually or commonly occurring or done 不平常的，异常的
6. rare [reə(r)] *adj.* not occurring very often 稀罕的，罕见的
7. discourage [dɪsˈkʌrɪdʒ] *vt.* to try to prevent sth. or to prevent sb. from doing sth. 阻挡，阻止
8. current [ˈkʌrənt] *adj.* belonging to the present time; happening or being used or done now 目前的
9. advantage [ədˈvɑːntɪdʒ] *n.* something puts you in a better position than other people 优势，有利条件
10. executive [igˈzekjətɪv] *n.* a person with senior managerial responsibility in a business 经理；主管领导

☞Phrases and Expressions

1. the auto business: the car industry 汽车行业
2. self-driving vehicle: the car capable of travelling without input from a human operator, by means of computer systems working in conjunction with on-board sensors 自动驾驶汽车
3. in theory: used in describing what is supposed to happen or be possible, usually with the implication that it does not in fact happen 理论上讲
4. let alone: used to indicate that something is far less likely or suitable than something else already mentioned 更不用说；更别提

☞Notes

1. BMW: Bavarian Motor Work (BMW) is a world-famous luxury automotive brand. Founded in 1916, BMW was headquartered in Munich, Bavaria State, Germany. The logo in blue-and-white of BMW implies the state flag of Bavaria, where the brand was founded. In 2015, BMW Group ranked first in the global high-end automotive market for 11 consecutive years, which also meant it has set the new sales record for 5 consecutive years.

宝马是享誉世界的豪华汽车品牌。创建于 1916 年，总部设在德国巴伐利亚州慕尼黑。宝马的蓝白标志取自宝马总部所在地巴伐利亚州州旗的颜色。2015 年，宝马集团连续第 11 年问鼎全球高档车市场，也意味着其连续五年创下销售纪录。

2. Volkswagen: Volkswagen, the core of Volkswagen Group which is one of the four largest automotive producers around the world, is an automotive producer headquartered in Wolfsburg, Germany. 大众汽车是一家总部位于德国沃尔夫斯堡的汽车制造公司，也是世界四大汽车生产商之一的大众集团的核心企业。
3. Tesla: Tesla, headquartered in Silicon Valley, California, America, is an American electric automotive producer and energy supplier engaging in producing and selling electric automotive, solar panels and energy storage devices. 特斯拉是一家美国电动车及能源公司，产销电动车、太阳能板、储能设备。总部位于美国加利福尼亚州硅谷。

☞Exercises

Vocabulary

Fill in the blanks in each sentence by selecting the most suitable words from the word bank. Change the form when necessary.

positive	opportunity	communicate	continual	reward	minimum	commitment
favorite	post	virtual	access	benefit	gap	embarrass

1. The years he spent in the countryside proved to be a(n) ____ experience.
2. You can learn a lot from this online course. It is designed to help people ____ better through speech and writing.
3. Over a third of the population was estimated to have no ____ to health service.
4. Asking too many personal questions during an interview can lead to a(n) ____ situation.
5. Don't just complain about what's wrong with it; suggest some ____ ways to solve the problem.
6. No one in the class could match John's hard work and ____ to study, which is way the professor liked him.
7. The website allows you to take a(n) ____ tour of the well-known city which was about 2,000 years ago.
8. Those who dare not answer questions in a traditional classroom can easily enjoy the ____ of online courses.
9. He couldn't join the police because he was below the ____ height allowed by the rules.
10. Many new ____ will be opened up in the future for those with a university education.

Unit 4　Aeronautics and Space

✧ Text A

The International Space Station's Evening Show

Hello and welcome! I'm Jim Tedder in Washington with the program that helps you learn and improve your American English. Today we have something special for you. We will meet with some friends not far from the VOA studios and all together look up at the night sky in search of a wonderful machine, far above us.

Americans often watch television or go to movies or parties on weekends. But on a recent Saturday night, a few people met for unusual activity at a public park in Bethesda, Maryland. A neighbor had invited them to watch the International Space Station (ISS) pass overhead.

The park had few lights to compete with stars and other objects in the sky. That helped make it possible to find the International Space Station just by looking up. No special equipment was needed.

At the time, the space station was carrying six crew members. They represented the space programs in Russia, Japan and the United States. Three of the men returned safely to Earth not long after the sighting in Maryland.

Oleg Kotov commanded the return trip to Kazakhstan. He is a pilot, an officer in the Russian Air Force and a medical doctor who was born in Crimea. With him were Russian cosmonaut Sergei Ryazanskiy and American astronaut Mike Hopkins.

Three other men stayed on the ISS. Three others joined them after leaving the Baikonur Cosmodrome on March 25th.

Back in Maryland, the space station watchers gathered among children's swings and toy cars. They sought to identify the stars and planets in the clear night sky. One woman repeatedly looked at her wristwatch.

"Hey, there it is," called a man. He pointed at yellow, red and blue lights sailing through the night. But no, that was an airplane.

"Isn't that it," cried another watcher? She caught sight of a lighted object as it moved through the skies. No, that also was a plane. The woman worried that perhaps they had

missed the sighting.

Luckily, help was available. Retired television engineer Paul Monte-Bovi served as a kind of guide and host for the Saturday night event. He has belonged for many years to one of America's largest amateur astronomy clubs.

Mr. Monte-Bovi owns three telescopes. He noted that the space station might look as if it were on the same level with airplanes. But he said it is really more than 300 kilometers up in the sky.

And its light looks bright white. The ISS does not show up as the yellowish color of the stars and other objects in the sky. "It does look like an airplane, the difference being that the space station does not have flashing lights."

The Maryland sky watchers found the ISS by looking to the northwest and watching it move southeast. NASA, the American space agency, says that it is not usually the case. But the space station's trajectories, the curved paths, differ.

As promised, the station appeared at the predicted time. That night, it was first observed at 7:44 p.m. Eastern Daylight Time, and seen for about four minutes.

"It is way high, and it's traveling in a continuous move."

Paul Monte-Bovi said what the people were seeing was the reflection of the sunlight on the station's solar panels. This equipment produces the many kilowatts of power required to keep the space station operating. And it seems that the bright dot in the sky is huge. The panels alone cover four tenths of one hectare.

One of the skywatchers asked the others to predict what the crew members were doing overhead. That question remained unanswered. But people could find out what the space station would do in the coming week.

For example, on the following Tuesday and Thursday, the station deployed micro-satellites known as NanoRacksCubeSats. These devices are supposed to increase scientific observation of the Earth. Another goal for the ISS is to learn the effects of space on human bodies. Crew members tested those effects on their mental abilities and organs including the heart and bones.

Another experiment is meant to increase our understanding of protein structure and how proteins operate. Improved understanding of proteins might someday provide treatment for several currently incurable conditions.

It was a comet, not a space station, that got Paul Monte-Bovi interested in amateur astronomy in 1997. The comet Hale-Bopp had been discovered just two years earlier. Some people say it was the brightest comet in history. He remembered the event.

"I heard about comets but never thought I would be able to see one. And a neighbor came by and said 'Let's go see the comet. I hear it's out tonight.' And I had a breathtaking view."

That experience led him to join the Northern Virginia Astronomy Club. Recently, another club member told him that NASA provides information about the times and places when people can see the ISS.

Mr. Monte-Bovi is among about 85,000 people who have asked for that information. Now he asks neighbors to join him for sightings when he thinks the ISS would be easiest to see.

"We have neighbors who have some young children who are real space cadets (fans). They love anything to do with the night sky... I thought it would be fun to have the rest of the neighborhood become aware of it..."

NASA's Mission Control provides the details about where it is possible to see the International Space Station in 4,600 places around the world. You can ask to receive an e-mail telling when you can expect to see the ISS near home. If your community is not on the list, choose a place that is close. For more information, go to spot the station.

When we are not looking up at the night sky, we Americans often spend our free time reading books. According to the *New York Times* newspaper, the most popular fiction book in the United States is *Missing You*. It was written by Harlan Coben. It is the story of Kat Donovan, a New York police detective who searches for the lover who left her years before.

(Adapted from *Aeronautics and Space Flight Collections*. 2007, pp. 23-25. by Peter Daniel)

☞New Words

1. crew [kruː] *n.* all the people working on a ship, plane, etc. 全体船员；全体乘务员；一群，一帮
2. represent [ˌreprɪˈzent] 1) *vt.* to act or speak officially for sb. and defend their interests 表达，表示，展现 2) *vt.* to express or complain about something, to a person in authority 代表，代理
3. amateur [ˈæmətə(r)] *n.* a person who takes part in a sport or other activity for enjoyment, not as a job 业余爱好者；业余运动员
4. astronomy [əˈstrɒnəmi] *n.* the scientific study of the sun, moon, stars, planets, etc. 天文学
5. continuous [kənˈtɪnjuəs] *adj.* happening or existing for a period of time without interruption 不断的；持续的；连续的
6. panel [ˈpænl] *n.* a square or rectangular piece of wood, glass or metal that forms part of a larger surface such as a door or wall 嵌板，镶板，方格板块
7. cosmonaut [ˈkɒzmənɔːt] *n.* an astronaut from Soviet Union （苏联的）宇航员，航天员，太空人
8. identify [aɪˈdentɪfaɪ] *vt.* to recognize sb./sth. and be able to say who or what they are 确认；认出；鉴定

9. device [dɪˈvaɪs] *n.* an object or a piece of equipment that has been designed to do a particular job 装置；仪器；器具；设备
10. protein [ˈprəʊtiːn] *n.* a substance, found within all living things, that forms the structure of muscles, organs, etc. There are many different proteins and they are an essential part of what humans and animals eat to help them grow and stay healthy. 蛋白质
11. incurable [ɪnˈkjʊərəbl] *adj.* not able to be cured 不能治愈的
12. aware [əˈweə(r)] *adj.* knowing or realizing sth. 知道；意识到；明白
13. detective [dɪˈtektɪv] *n.* a person, especially a police officer, whose occupation is to investigate and solve crimes 侦探
14. kidnap [ˈkɪdnæp] *v.* to take sb. away illegally and keep them as a prisoner, especially in order to get money or sth. else for returning them 绑架；诱拐；拐骗
15. award-winning [əˈwɔːdwɪnɪŋ] *adj.* having received awards 成功的，优等的，一流的
16. slavery [ˈsleɪvəri] *n.* the state of being a slave 奴隶制度；奴隶身份；苦役，奴隶般的劳动；奴役，束缚，屈从，耽迷（酒色等）
17. escape [ɪˈskeɪp] *v.* to get away from a place where you have been kept as a prisoner or not allowed to leave （从监禁或管制中）逃跑，逃走，逃出
18. bondage [ˈbɒndɪdʒ] *n.* the state of being a slave or prisoner 奴役；束缚
19. broadcast [ˈbrɔːdkɑːst] *v.* to send out programs on television or radio 播送（电视或无线电节目）；广播

☞Phrases and Expressions

1. serve as: to act as 充当
2. be supposed to: to be ought to, should 应该
3. test effects on: to have an impact on 对……有影响
4. move aside: to put sth. …aside 除去；搁到旁边
5. at the beginning of: at first 首先；起初

☞Notes

1. ISS: the International Space Station (ISS) is the largest international scientific project in human history. Construction on the ISS began in 1998 and was completed in 2002. 国际空间站是人类历史上最大的国际科学项目。国际空间站的建设始于 1998 年，于 2002 年完成。
2. NASA: The National Aeronautics and Space Administration (NASA/ˈnæsə/) is an independent agency of the United States Federal Government responsible for the civilian space program, as well as aeronautics and aerospace research. 美国国家航空航天局是美

国联邦政府的一个独立机构，负责民用空间计划以及航空和航天研究。

3. solar panels: solar panels absorb sunlight as a source of energy to generate electricity. A photovoltaic (PV) module is a packaged, connected assembly of typically 6×10 photovoltaic solar cells. Photovoltaic modules constitute the photovoltaic array of a photovoltaic system that generates and supplies solar electricity in commercial and residential applications. 太阳能电池板吸收太阳光作为发电的能源。光伏模块是通常为 6×10 的光伏太阳能电池的封装的连接组件。光伏模块构成光伏系统的光伏阵列，其在商业和住宅应用中产生并提供太阳能电力。

☞Exercises

Ⅰ. Reading Comprehension

Answer the following questions according to the text.

1. What were the lights used for?
2. How many people was the space station carrying?
3. Who is Oleg Kotov?
4. When did the other man leave the Baikonur Cosmodrome?
5. How can the watchers find the ISS?

Ⅱ. Vocabulary

A. Fill in the blanks in the paragraph by selecting the most suitable words from the word bank.

A. connected	B. emerge	C. calculated	D. extend	E. unprecedented	F. annual
G. postpone	H. Thanks	I. rapidly	J. increasingly	K. approve	L. convergence
M. distinct	N. edge	O. enabler			

Connectivity is the foundation for the digital economy. The Internet has already __1__ more than three billion users across the globe and about 14 billion devices. A major challenge is how to __2__ connectivity not only to the next several billion users, but also the next 50 billion devices. __3__ to the Internet of Things, the economy is digitalizing on an __4__ scale, as devices and objects connect to the Internet's network of networks and communicate with each other. A world is emerging where smart sensors and actuators will __5__ monitor the health, location and activities of people and animals, the state of the natural environment, the quality of food and more. If connectivity is the foundation and __6__, the driver that makes all this possible is convergence. Thanks to digitalization and the growing capabilities of the Internet, there has been an ongoing __7__ between once __8__ parts of

 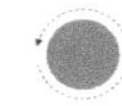

the communication ecosystem. Fixed and wireless networks converge, voice and text, and telecommunication and broadcasting too. Your car, your home appliances: all are being connected. Companies best known for communication equipment increasingly provide content, and vice-versa. Each firm may have a competitive __9__ and niche function but old lines are blurring as new services and possibilities __10__ all the time. Much of the time you will not be aware that the tram or stop you are waiting at is connected to the Internet, but they will have IP addresses.

B. Fill in the blanks in each sentence by selecting the most suitable words.

1. Most importantly, such an experience helps ____ a heightened sensitivity to other cultures and will bring about a greater appreciation of one's own culture as well.
   A. coach　　B. forsake　　C. foster　　D. censor
2. When Ann broke the dish she tried to put the ____ back together.
   A. fragments　　B. pieces　　C. bits　　D. slices
3. Jane tried to ____ the doorman with money, but she failed.
   A. bribe　　B. corrupt　　C. award　　D. endow
4. Classification is a useful ____ to the organization of knowledge in any field.
   A. means　　B. approach　　C. mode　　D. manner
5. The human race has already paid a heavy price for its slow ____ to environmental threats.
   A. response　　B. responsibility　　C. resolution　　D. resistance
6. We have high regard for Prof. Joseph because he always ____ his principles.
   A. lives on　　B. lives up to　　C. lives through　　D. lives with
7. My grandfather accidentally ____ fire to the house.
   A. put　　B. took　　C. set　　D. got
8. We enjoyed the holiday ____ the expense.
   A. except　　B. besides　　C. in addition to　　D. except for
9. If you want children to work hard you must ____ their interests instead of their sense of duty.
   A. appeal to　　B. look into　　C. give rise to　　D. go in for
10. Basically, a robot is a machine that moves, manipulates, joins or processes ____ in the same way as human hand or arm.
   A. characters　　B. components　　C. catalogues　　D. collections

Ⅲ. Cloze

Choose an appropriate word from the four choices marked A, B, C and D for each blank in the passage.

Music produces profound and lasting changes in the brain. Schools should add music classes, not cut them. Nearly 20 years ago, a small study advanced the __1__ that listening to Mozart's Sonata for Two Pianos in D Major could boost mental functioning. It was not long __2__ trademarked "Mozart effect" products began to appeal to anxious parents aiming to put toddlers (刚学步的孩子) __3__ the fast track to prestigious universities like Harvard and Yale. Georgia's governor even __4__ giving every newborn there a classical CD or cassette.

The __5__ for Mozart therapy turned out to be weak, perhaps nonexistent, although the __6__ study never claimed anything more than a temporary and limited effect. In recent years, __7__, scientists have examined the benefits of a concerted __8__ to study and practice music, as __9__ to playing a Mozart CD or a computer-based "brain fitness" game __10__ in a while.

Advanced monitoring __11__ have enabled scientists to see what happens __12__ your head when you listen to your mother and actually practice the violin for an hour every afternoon. And they have found that music __13__ can produce profound and lasting changes that __14__ the general ability to learn. These results should __15__ public officials that music classes are not mere decoration, ripe for discarding in the budget crises that constantly __16__ public schools.

Studies have shown that __17__ instrument training from an early age can help the brain to __18__ sounds better, making it easier to stay focused when absorbing other subjects, from literature to mathematics. The musically adept (擅长的) are better able to __19__ on a biology lesson despite the noise in the classroom __20__, a few years later, to finish a call with a client when a colleague in the next office starts screaming a subordinate. They can attend to several things at once in the mental scratch pad called working memory, an essential skill in this era of multitasking.

1. A. notice	B. note	C. notion	D. notification
2. A. that	B. until	C. since	D. before
3. A. up	B. by	C. on	D. at
4. A. propelled	B. proposed	C. submitted	D. subjected
5. A. witness	B. evidence	C. symptom	D. context
6. A. subtle	B. elementary	C. sensitive	D. original
7. A. however	B. moreover	C. then	D. therefore
8. A. effort	B. impulse	C. object	D. attention
9. A. opposed	B. accustomed	C. related	D. devoted
10. A. quite	B. once	C. often	D. much
11. A. organisms	B. techniques	C. mechanisms	D. mechanics
12. A. upon	B. amid	C. among	D. inside
13. A. subjects	B. models	C. causes	D. lessons
14. A. enhance	B. introduce	C. accelerate	D. elaborate

15. A. contend	B. convey	C. conceive	D. convince
16. A. trouble	B. transform	C. distract	D. disclose
17. A. urgent	B. casual	C. diligent	D. solemn
18. A. proceed	B. process	C. prefer	D. predict
19. A. count	B. concentrate	C. insist	D. depend
20. A. but	B. or	C. for	D. so

Ⅳ. Translation

Translate the following text into Chinese.

When you're flying over Nepal, it's easy to soar in your imagination and pretend you're tiny—a butterfly—and drifting above one of those three-dimensional topographical maps architects use, the circling contour lines replaced by the terraced rice paddies that surround each high ridge.

Nepal is a small country, and from the windows of our plane floating eastward at 12,000 feet, one can see clearly the brilliant white mirage of the high Himalayas thirty miles off the left window.

Out of the right window, the view is of three or four high terraced ridges giving sudden way to the plains of India beyond.

There were few roads visible below, most transportation in Nepal being by foot along ancient trails that connect and bind the country together. There is also a network of dirt airstrips, which was fortunate for me, as I had no time for the two-and-a-half week trek to my destination. I was on a flight to the local airport.

Ⅴ. Writing

Write an essay on the following topic with no less than 150 words. Give reasons for your answer and include any relevant examples from your own knowledge or experience.

The Importance of Waste Sorting

✧ Text B

Planetary Protection Officer and New Findings

NASA Is Looking for Planetary Protection Officer. Does this job sound like it might be for you? The job opening was announced by the United States space agency, NASA. The Washington Post called the job title "one of the greatest ever conceived". If you think this job seems just right for action movie stars like Bruce Willis in "Armageddon" or Will Smith in

"Men in Black" or "Independence Day", you are not alone.

The announcement says the job pays up to $187,000 yearly. It is getting a lot of attention on social media this week. One person wondered if the job comes with a cape like a superhero would wear. But before you start practicing your fighting skills, take a closer look at the description of the job: "Planetary protection is concerned with the avoidance of... biological contamination in human and robotic space exploration."

This means the person who gets this job will be in charge of making sure Earth organisms are not accidentally taken to other planets. He or she will also be responsible for making sure samples collected on other planets are kept safe when they make it back to Earth.

NASA saw the job listing was getting a lot of attention, so it posted a video from Catharine Conley, the current officer.

The video explained the job by comparing what happens on Earth when non-native species are brought to a new continent.

When plants, fish, insects and other animals come to an area where they have no natural predators, they can kill native plants, eat other animals and damage crops. NASA is concerned that might happen if a space vehicle brings something from Earth to another planet. Conley is a biologist. So if your science background is not strong, you might not be a good fit for the job. She has not yet said if she will re-apply for the job.

If you still think you might be a good fit for the job, keep one other thing in mind: you must be a U.S. citizen. National Aeronautics and Space Administration scientists announced on Wednesday the discovery of 715 new planets around distant stars, including four alien worlds roughly the size of Earth that might be the proper temperature for liquid water to form and, therefore, potentially suitable for life.

This latest discovery, based on two years of data collected from 150,000 or so stars by the agency's orbiting Kepler space telescope, brings the confirmed count of planets outside our solar system to nearly 1,700 worlds.

We have almost doubled the number of planets known to humanity, said planetary scientist Jack Lissauer at NASA's Ames Research Center in Mountain View, Calif., who is a science co-investigator on the $600 million Kepler space telescope mission.

The team of three dozen astronomers, data analysts and planetary scientists detailed their findings in two research papers submitted to *The Astrophysical Journal* and discussed their findings on Wednesday during a press conference held by NASA.

Almost all of these newly verified exoplanets—as the alien worlds are called—are smaller than the planet Neptune, a gaseous giant at the outer reaches of our solar system that is almost four times the size of Earth. The worlds are clustered around just 305 stars in solar systems that, like our own, contain multiple planets, the scientists said.

The researchers said that four of these newly confirmed planets are less than 2.5 times the size of Earth and orbit in the so-called habitable zone around their stars—that is, the distance at which the surface temperature of an orbiting planet may be right for liquid water. That means it would be not so hot that it would boil into space and not so cold that it would freeze solid.

One of those new planets, called Kepler-296f, is twice the size of Earth and orbits a star half the size and only 5% as bright as our sun, said Jason Rowe, a research scientist at the SETI Institute and a member of the Kepler science group.

Details of the others—designated Kepler 174d, Kepler 298d and Kepler 309c—weren't publicly available on Wednesday. The total, though, increases the number of Earth-sized planets by 400%, Mr. Rowe said.

Taken together, the new Kepler discoveries confirm that small planets are extremely common in our galaxy, said planetary physicist Sara Seager at the Massachusetts Institute of Technology, who wasn't involved in the discoveries. I am extremely excited about this.

The data were collected before the Kepler telescope malfunctioned last year, leaving it unable to track stars precisely enough to continue the planet-hunting mission for which it was launched in 2009. The Kepler scientists are now seeking funding from NASA to operate the telescope for another two years, in a reduced role, to study how planets form around stars.

In the meantime, the researchers continue to pore through data collected while it was still working properly. Several thousand possible planet candidates exist, of which scientists expect to be able to confirm several hundred more planets in orbit around other stars.

(From *Bibliography of Aeronautics*. 2017, 3. pp. 41-43. by John Maitri)

☞New Words

1. species [ˈspiːʃiːz] *n.* a group into which animals, plants, etc. that are able to breed with each other 物种；种类；类型
2. planetary [ˈplænətri] *adj.* relating to a planet or planets 行星的
3. avoidance [əˈvɔɪdəns] *n.* not doing sth.; preventing sth. from existing or happening 避免；防止；回避；避开
4. description [dɪˈskrɪpʃn] *n.* a piece of writing or speech that says what sb./sth. is like; the act of writing or saying in words what sb./sth. is like 描述；形容；种类；类型
5. contamination [kənˌtæmɪˈneɪʃn] *n.* the act of contaminating or polluting 污染；弄脏；毒害；玷污
6. current [ˈkʌrənt] *adj.* happening now; of the present time 现在的；最近的；流行的；流传的

7. announce [əˈnaʊns] *v.* to make a formal public statement about a fact, occurrence, or intention 宣布；宣告；公布
8. exploration [ˌekspləˈreɪʃn] *n.* the act of travelling through a place in order to find out about it or look for sth. in it 探测，勘探，探险；搜索，研究
9. previously [ˈpriːviəsli] *adv.* at some time before the period that you are talking about 以前；事先；仓促
10. announcement [əˈnaʊnsmənt] *n.* a spoken or written statement that informs people about sth. 公告，布告，通告
11. orbit [ˈɔːbɪt] *n.* a curved path followed by a planet or an object as it moves around another planet, star, moon, etc.（天体等运行的）轨道
12. alien [ˈeɪliən] *adj.* strange and frightening; different from what you are used to (to sb./sth.) 陌生的；不熟悉
13. mission [ˈmɪʃn] *n.* an important official job that a person or group of people is given to do, especially when they are sent to another country 官方使命；使团的使命
14. detail [ˈdiːteɪl] *n.* a small individual fact or item; a less important fact or item 细微之处；枝节；琐事
15. submit [səbˈmɪt] *v.* to give a document, proposal, etc. to sb. in authority so that they can study or consider it 提交，呈递（文件、建议等）
16. malfunction [ˌmælˈfʌŋkʃn] *n.* (of a piece of equipment or machinery) to fail to function normally（机器）失灵，出故障；（人体器官）机能失常，出现功能障碍
17. reduce [rɪˈdjuːs] *v.* to make sth. less or smaller in size, quantity, price, etc. 减少，缩小（尺寸、数量、价格等）
18. properly [ˈprɒpəli] *adv.* in a way that is correct and/or appropriate 正确地；适当地
19. candidate [ˈkændɪdət] *n.* a person who is trying to be elected or is applying for a job (for sth.) （竞选或求职的）候选人，申请人
20. confirm [kənˈfɜːm] *v.* to state or show that sth. is definitely true or correct, especially by providing evidence （尤指提供证据来）证实，证明，确认

☞Phrases and Expressions

1. be fit for: to be suitable for 适合……
2. above ground: at or above ground level 地上
3. …get a lot of attention: to pay much attention to… 受到广泛关注
4. keep in mind: to bear in mind, know sth. by heart 牢记于心

Notes

1. *The Washington Post*: *The Washington Post* is a major American daily newspaper founded on December 6, 1877. It is the largest newspaper published in Washington, D.C., the capital city of the United States, (and 8th in the US) and has a particular emphasis on national politics. Its slogan "Democracy Dies in Darkness" appears on its masthead. Daily broadsheet editions are printed for the District of Columbia, Maryland, and Virginia. 《华盛顿邮报》是 1877 年 12 月 6 日成立的美国主要日报。它是美国首都华盛顿特区发行的最大（美国第八大）报纸，特别强调国家政治。其标头上印有“民主在黑暗中死亡”的标语。在哥伦比亚特区，马里兰州和弗吉尼亚州发行每日宽幅版。
2. Planetary Protection: Planetary protection is a guiding principle in the design of an interplanetary mission, aiming to prevent biological contamination of both the target celestial body and the Earth in the case of sample-return missions. Planetary protection reflects both the unknown nature of the space environment and the desire of the scientific community to preserve the pristine nature of celestial bodies until they can be studied in detail. There are two types of interplanetary contamination. Forward contamination is the transfer of viable organisms from Earth to another celestial body. Back contamination is the transfer of extraterrestrial organisms, if such exists, back to the Earth's biosphere. 行星保护是设计星际任务的指导原则，旨在防止目标天体和地球在返回样本时的生物污染。行星保护既反映了空间环境的未知性质，也说明科学界期望在能够对天体进行详细研究之前，保持它们的原始性质。有两种类型的星际污染。正向污染是将有生命的有机体从地球转移到另一个天体。返回污染是外星生物（如果存在的话）转移回地球的生物圈。
3. Kepler: Kepler is a space observatory launched by NASA to discover Earth-size planets orbiting other stars. Named after astronomer Johannes Kepler, the spacecraft was launched on March 7, 2009, into an Earth-trailing heliocentric orbit. The principal investigator was William J. Borucki. Designed to survey a portion of our region of the Milky Way to discover Earth-size exoplanets in or near habitable zones and estimate how many of the billions of stars in the Milky Way have such planets, Kepler's sole scientific instrument is a photometer that continually monitors the brightness of approx 150,000 main sequence stars in a fixed field of view. These data are transmitted to Earth, then analyzed to detect periodic dimming caused by exoplanets that cross in front of their host star. 开普勒是由美国国家航空航天局发射的一个太空观测站，用于探测围绕其他恒星运行的地球大小的行星。该飞船以天文学家约翰内斯·开普勒的名字命名，于 2009 年 3 月 7 日发射到地球尾随的日心轨道。首席研究员是威廉·J. 伯鲁奇。开普勒唯一的科学仪器是光度计，能持续监测固定视野内约 150,000 个主序星的亮度。其设计初衷是探测人类在银河系中所处区域的某一部分，希望能发现可生存区域当中或附近与地

球大小相当的系外行星，同时估测银河系的数十亿颗恒星拥有多少这样的行星。这些数据被传输到地球，分析后用以检测由在主星前面穿过的系外行星引起的周期性变暗。

☞Exercises

Vocabulary

Fill in the blanks in the paragraph by selecting the most suitable words from the word bank.

A. scale	B. retailed	C. generate	D. extreme	E. technically	F. affordable
G. situation	H. really	I. potential	J. gap	K. voluntary	L. excessive	M. insulted
N. purchase	O. primarily				

The flood of women into the job market boosted economic growth and changed U.S. society in many ways. Many in-home jobs that used to be done __1__ by women—ranging from family shopping to preparing meals to doing __2__ work—still need to be done by someone. Husbands and children now do some of these jobs, a __3__ that has changed the target market for many products. Or a working woman may face crushing "poverty of time" and look for help elsewhere, creating opportunities for producers of frozen meals, child care centers, dry cleaners, financial services, and the like.

Although there is still a big wage __4__ between men and women, the income working women __5__ gives them new independence and buying power. For example, women now __6__ about half of all cars. Not long ago, many car dealers __7__ women shoppers by ignoring them or suggesting that they come back with their husbands. Now car companies have realized that women are __8__ customers. It's interesting that some leading Japanese car dealers were the first to __9__ pay attention to women customers. In Japan, fewer women have jobs or buy cars—Japanese society is still very much male-oriented. Perhaps it was the __10__ contrast with Japanese society that prompted American firms to pay more attention to women buyers.

Unit 5 Material Engineering

✧ Text A

Will 3-D Printing Transform Conventional Manufacturing?

Oak Ridge National Laboratory's Robotic prosthesis looks like something out of medieval times—a hand clad in chain mail more appropriate for wielding a broadsword than a mug of coffee. Both the underlying skeleton and thin, meshlike skin are made of titanium to make the hand durable and dexterous while also keeping it lightweight. The powerful miniature hydraulics that move the fingers rely on a network of ducts integrated into the prosthesis's structure—no drilled holes, hoses or couplings required.

Yet what makes this robot hand special is not what it can make or do but rather how it was made and what it represents. Conceived on a computer and assembled from a few dozen printed parts by so-called additive manufacturing, more popularly known as 3-D printing, Oak Ridge's invention offers a glimpse into the future of manufacturing—a future where previously impossible designs can be printed to order in a matter of hours.

"You're looking at a very, very complex design that has internal hydraulic tubing that can be run in excess of 3,000 pounds per square inch," says Craig Blue, director of Oak Ridge's energy materials program. "You have meshing to make it a lightweight structure, putting material only where you need it. There's no technology today, other than additive manufacturing, that can make that *robotic hand*."

As 3-D printing matures to the point where it can make complex machinery that can't be made any other way, big-volume manufacturers such as Boeing and GE are starting to apply the technology to their advanced product lines. Instead of the old approach of carving a usable part out of a large block of material, additive manufacturing builds an object up layer by layer. This shift in thinking has the potential to affect every facet of manufacturing—from prototype design to mass-produced product.

Yet technical challenges continue to be devil 3-D printing. Compared with ordinary subtractive manufacturing, additive manufacturing can be slow, the fit and finish of its materials inconsistent. Further, 3-D printers have trouble building objects out of multiple kinds of materials, and they cannot yet integrate electronics without frying the circuits.

Researchers are working hard to overcome these limitations—and few doubt that for customizable, small-volume applications, additive manufacturing has tremendous power. As technology expands into the mass marketplace, 3-D printing could begin to power a widespread manufacturing revolution.

Additive Advances

The origins of 3-D printing stretch back to the late 1980s, when start-up companies and academics—most notably at the University of Texas at Austin—invented machines that could build three-dimensional models of digital designs in minutes. For decades those systems and similar types, which first cost around $175,000 gained notoriety for their ability to help inventors and engineers rapidly and relatively inexpensively produce their prototypes.

Since then, 3-D printing has taken two paths. At one extreme, hobbyists and would-be entrepreneurs can whip up plastic models using machines that cost $2,000 and less. These kitchen counter devices allow users to invent new objects—a technology that has invited comparisons between 3-D printing and personal computers. "In the same way, the Internet, the cloud and open-source software have allowed small teams to live on ramen noodles for six months and build an app, post it and see if anyone is interested, we're beginning to see that same phenomenon spread to manufactured goods," says Tom Kalil, deputy director of technology and innovation at the White House's Office of Science and Technology Policy.

At the other extreme, large manufacturers are cultivating advanced, industrial-strength approaches to produce aircraft parts and biomedical devices such as replacement hips. The machines required to do this cost upward of $30,000 with laser-based appliances that make high-quality metal products selling for as much as $1 million. These printers can use polymers, metals or other materials in liquid or powder form. Objects begin as digital files, enabling designers to tweak their work before the building begins, with little impact on cost.

Printing in 3-D could replace certain conventional mass-production processes such as casting, molding and machining by 2030, especially in the case of short production runs or manufacturers aiming for more customized products, according to the "Global Trends 2030: Alternative Worlds" report issued last November by the National Intelligence Council, a team of analysts supporting the Office of the Director of National Intelligence. Aerospace companies are at the forefront of this trend. GE Aviation, which has been making aircraft engines for nearly a century, recently bought two suppliers that specialize in making aircraft parts via additive manufacturing processes. Boeing already uses 3-D printing to make more than 22,000 parts used on its civilian and military aircraft.

Such companies are discovering that 3-D printing can also be more efficient than conventional production, both in terms of energy and materials. "If you're machining a part, it's not unusual that 80 to 90 percent of the block of material you start with can end up as chips or scraps on the floor," says Terry Wohlers, principal consultant and president of Wohlers Associates, an additive manufacturing consulting firm in Fort Collins, Colo.

(From *Scientific American*, 2013, 5. pp. 46-47. by Greenemeier Leech)

☞New Words

1. prosthesis [prɒsˈθiːsɪs] *n.* corrective consisting of a replacement for a part of the body 假体；假肢
2. medieval [ˌmediˈiːvl] *adj.* relating to or belonging to the Middle Ages 中世纪的；原始的；仿中世纪的；老式的
3. clad [klæd] *adj.* wearing or provided with clothing; sometimes used in combination 覆盖的
4. titanium [tɪˈteɪniəm] *n.* a light strong grey lustrous corrosion-resistant metallic element used in strong lightweight alloys (as for airplane parts) 钛（金属元素）
5. durable [ˈdjʊərəbl] *adj.* serviceable for a long time 耐用的，持久的
6. dexterous [ˈdekstrəs] *adj.* skillful in physical movements; especially of the hands 灵巧的；敏捷的
7. hydraulics [haɪˈdrɒlɪks] *n.* study of the mechanics of fluids 水力学
8. drill [drɪl] *v.* to make a hole with a special tool 钻孔，打钻
9. conceive [kənˈsiːv] *v.* to form an idea, a plan, etc. in your mind; to imagine sth. 想出（计划等）；想象；构想；设想
10. assemble [əˈsembl] *vt.* to fit together the separate component parts of (a machine or other object) 装配
11. glimpse [glɪmps] *n.* a brief or incomplete view 一瞥；一看
12. mesh [meʃ] *n.* an interlaced structure 啮合；建网
13. additive [ˈædətɪv] *adj.* characterized or produced by addition 附加的
14. prototype [ˈprəʊtətaɪp] *n.* a standard or typical example 原型；标准，模范
15. devil [ˈdevl] *n.* (in Christian and Jewish belief) the chief evil spirit; Satan 魔鬼；撒旦
16. subtract [səbˈtrækt] *v.* to take a number or an amount away from another number or amount 减，减去
17. fry [fraɪ] *v.* to cook on a hot surface using fat 油炸；油煎
18. biomedical [ˌbaɪəʊˈmedɪkl] *adj.* relating to the activities and applications of science to clinical medicine 生物医学的
19. casting [ˈkɑːstɪŋ] *n.* an object formed by a mold 铸造；铸件
20. aerospace [ˈeərəʊspeɪs] *n.* the atmosphere and outer space considered as a whole 航空宇宙
21. scrap [skræp] *n.* a small fragment of something broken off from the whole 碎片；残余物

☞Phrases and Expressions

1. rely on: to be dependent on, as for support or maintenance 依赖于
2. whip up: to prepare or cook quickly or hastily 激起
3. three-dimensional: involving or relating to three dimensions or aspects; giving the illusion of depth 三维的；立体的
4. apply to: to concern or relate to sb./sth. 涉及，与……有关
5. at the forefront of: at the newest and most exciting stage in the development of something 处在最前沿、最先进的
6. specialize in: to concentrate on and become expert in a particular subject or skill 专业从事

☞Notes

1. 3-D printing: 3-D printing is any of various processes in which material is joined or solidified under computer control to create a three-dimensional object, with the material being added together (such as liquid molecules or powder grains being fused together). 3-D printing is used in both rapid prototyping and additive manufacturing. 3D 打印指在计算机控制下，任何将可黏合材料（如液体分子或粉末颗粒）以连接、固化等方式加工产生三维物体的过程，可用于快速打印与增材制造。
2. Oak Ridge National Laboratory (ORNL): an American multiprogram science and technology national laboratory sponsored by the U.S. Department of Energy (DOE) and administered, managed, and operated by UT-Battelle as a federally funded research and development center (FFRDC) under a contract with the DOE. 橡树岭国家实验室是隶属于美国能源部的一个大型国家级科技研究所。该所作为美国联邦资助的研发中心，由 Battelle 纪念研究所管理。

☞Exercises

Ⅰ. Reading Comprehension

Answer the following questions according to the text.

1. What does the robotic prosthesis in Oak Ridge National Laboratory look like?
2. What challenges does 3-D printing meet?
3. Does 3-D printing develop smoothly since its advent? Why?
4. What are the main spheres where 3-D printing is adapted?

Ⅱ. Vocabulary

A. Fill in the blanks in the paragraph by selecting the most suitable words from the word bank.

A. deficit	B. complained	C. severely	D. allowance	E. considerately	F. shuttle
G. evacuate	H. absently	I. adequate	J. dock	K. resume	L. vital	M. trivial
N. evaluate	O. fresh				

Two astronauts face a not-so-merry Christmas after being told to ration their food and hope a cargo ship with extra supplies docks on Dec. 21. Russian cosmonaut Salizhan Sharipov and American Leroy Chiao have been asked to cut out calories equal to three cans of Coke from their daily diet—around 10 percent of their daily __1__ and an amount that would be little noticed, NASA said. Russian officials, quoted in the local media, have __2__ blamed the previous crew for overeating during their one-month mission earlier this year, leaving a __3__ of meat and milk and a surplus of juice and confectionery. The Dec. 24 launch of the next Progress is now __4__ for the crew, stationed in orbit since October. It is due to __5__ with the ISS on Dec. 21. NASA officials said their situation was not so different from being cut off on Earth, and their lives were not at risk. If they do not receive __6__ supplies, the astronauts would have to __7__ the station and return to Earth on the Soyuz capsule that is docked there. Russia has been the sole lifeline to the ISS for almost two years when the United States grounded its __8__ fleet after the fatal Columbia accident. Russia has often __9__ of its financial struggle to keep the ISS fully serviced single-handedly. Shuttle flights could __10__ in May, officials have said, but in the meantime, Russia will continue to launch all manned and cargo ships.

B. Fill in the blanks in each sentence by selecting the most suitable words.

1. During the famine, many people were ____ to going without food for days.
   A. sunk　　B. reduced　　C. forced　　D. inclined
2. The computer can be programmed to ____ a whole variety of tasks.
   A. assign　　B. tackle　　C. realize　　D. solve
3. The team's efforts to score were ____ by the opposing goalkeeper.
   A. frustrated　　B. prevented　　C. discouraged　　D. accomplished
4. I only know the man by ____ but I have never spoken to him.
   A. chance　　B. heart　　C. sight　　D. experience
5. Being color-blind, Sally can't make a ____ between red and green.
   A. difference　　B. distinction　　C. comparison　　D. division
6. You must insist that students give a truthful answer ____ with the reality of their world.
   A. relevant　　B. simultaneous　　C. consistent　　D. practical

7. In order to raise money, Aunt Nicola had to ____ with some of her most treasured possessions.

A. divide B. separate C. part D. abandon.

8. The car was in good working ____ when I bought it a few months ago.

A. order B. form C. state D. circumstance

9. The customer expressed her ____ for that broad hat.

A. disapproval B. distaste C. dissatisfaction D. dismay

10. In order to repair barns, build fences, grow crops, and care for animals a farmer must indeed be ____ .

A. restless B. skilled C. strong D. versatile

Ⅲ. Cloze

Choose an appropriate word from the four choices marked A, B, C and D for each blank in the passage.

The translator must have an excellent, up-to-date knowledge of his source languages, full facility in the handling of his target language, which will be his mother tongue or language of habitual __1__ and a knowledge and understanding of the latest subject-matter in his field of specialization.

This is, as it were, his professional equipment. __2__ this, it is desirable that he should have an inquiring mind, wide interests, a good memory and the ability to grasp quickly the basic principles of new developments. He should be willing to work __3__ his own, often at high speeds, but should be humble enough to consult others __4__ his own knowledge not always prove adequate to the task in hand. He should be able to type fairly, quickly and accurately and, if he is working mainly for publication, should have more than a nodding __5__ with printing techniques and proof-reading. If he is working basically as an information translator, let us say, for an industrial firm, he should have the flexibility of mind to enable him to __6__ rapidly from one source language to another, as well as from one subject-matter to another, since this ability is frequently __7__ of him in such work. Bearing in mind the nature of the translator's work, i.e. the processing of the written word, it is, strictly speaking, __8__ that he should be able to speak the language he is dealing with. If he does speak them, it is an advantage __9__ a hindrance, but this skill is in many ways a luxury that he can __10__ with. It is, __11__, desirable that he should have an approximate idea about the pronunciation of his source languages even if this is restricted to __12__ how proper names and place names are pronounced. The same __13__ to an ability to write his source languages. If he can, well and good; if he cannot, it does not __14__. There are many other skills and __15__ that are desirable in a translator.

1. A. application B. use C. utility D. usage

2. A. More than	B. Except for	C. Because of	D. In addition to
3. A. of	B. by	C. for	D. on
4. A. should	B. when	C. because	D. if
5. A. familiarity	B. acquaintance	C. knowledge	D. skill
6. A. change	B. transform	C. turn	D. switch
7. A. lacked	B. required	C. faced	D. confronted
8. A. essential	B. unnecessary	C. advantageous	D. useless
9. A. over	B. despite	C. rather than	D. instead
10. A. deal	B. concern	C. work	D. do away
11. A. however	B. accordingly	C. consequently	D. thus
12. A. knowing	B. having known	C. know	D. have known
13. A. refers	B. comes	C. applies	D. amounts
14. A. matter	B. mind	C. harm	D. work
15. A. characteristics	B. qualities	C. distinctions	D. features

Ⅳ. Translation

Translate the following text into Chinese.

Both Boeing and Airbus have trumpeted the efficiency of their newest aircraft, the 787 and A350 respectively. Their clever designs and lightweight composites certainly make a difference. But a group of researchers at Stanford University, led by Ilan Kroo, has suggested that airlines could take a more naturalistic approach to cutting jet-fuel use, and it would not require them to buy new aircraft. The answer, says Dr Kroo, lies with birds. In 1914, a seminal paper by a German researcher called Carl Wieselsberger made scientists know that birds flying in formation—a V-shape, echelon, or otherwise—expend less energy. The air flowing over a bird's wings curls upwards behind the wingtips, a phenomenon known as upwash. Other birds flying in the upwash experience reduced drag, and spend less energy propelling themselves. Peter Lissaman, an aeronautics expert who was formerly at Caltech and the University of Southern California, has suggested that a formation of 25 birds might enjoy a range increase of 71%.

Ⅴ. Writing

Write an essay on the following topic with no less than 150 words. Give reasons for your answer and include any relevant examples from your own knowledge or experience.

How to Protect Intellectual Property

✧ Text B

My Boss the Robot

The minute Michael Dawson-Haggerty burst into my office, clad in a blackened lime-green welding jacket and wearing a big smile, I knew he and his partner had won. Their test: weld a metal space frame for a Humvee—a military vehicle ubiquitous in Iraq and Afghanistan—faster than a team of experts with decades of experience.

This was Dawson-Haggerty's first professional job—he had just completed his master's degree and joined the engineering staff at Carnegie Mellon University's Robotics Institute—and it is fair to say that he had been a little nervous as he got started. If truth be told, I was more worried about his partner, who was reliable enough but generally lacked people skills.

The cohort on this project was a robot, similar to those huge industrial machines we typically associate with assembly-line work at Ford or General Motors. Yet whereas those mechanical monsters operate inside cages to keep humans safely apart from unforgiving automated thrusts, we modified Spitfire—our 13 feet tall, one-armed welding robot equipped with a laser for an eye—to work right alongside a person. And instead of Spitfire taking orders from Dawson-Haggerty, the team tended to work the other way around: the robot dictated the next steps, with the hard work of positioning and welding divided between Dawson-Haggerty and Spitfire according to who could most efficiently complete the task. The robot, not the human, often called the shots. With the work so split, Dawson-Haggerty and his robot partner built the frame in 10 hours for $1,150, including raw materials and labor. The experts we had hired to serve as our control group performed the same task in 89 hours and billed us $7,075.

The economic consequences of a human's ability to work with a robot, and vice versa, are potentially enormous. Factories could do away with painstakingly configured assembly lines, saving billions in equipment setup costs. Need to modify a popular product? Human robot teams can create custom versions of anything from electronics to airplanes without the need for expensive retooling. The technology will allow companies to quickly respond to consumer demand, updating products in cycles measured in weeks, not years. And workers should find rewarding the ever-changing challenges of the factory floor.

For these reasons and more, we need to realize that robots may ultimately be more effective as supervisors, not slaves.

Keeping Your Head

There is always a lot of discussion surrounding what, exactly, a "robot" is. The robotics research community defines them as machines that can sense, think and act autonomously. This is not quite right—your house's thermostat can do all these things, yet you would not classify your house as a robot. The difference is that your thermostat is just a small part of what your house does.

Only when "robotic" functions are used in service of an object's core responsibility can the object itself be considered a robot. For example, when a self-driving car uses sensors and artificial intelligence to enable transportation—a car's essential function—it becomes a robot.

Manufacturers have deployed robots for more than half a century to improve efficiency through automation. Yet robots have been special purpose machines—excellent at, say, welding a certain set of joints on every car coming down an assembly line.

Humans have done the organization, setting up the assembly line to capitalize on their robots' strength and precision. The process works well for products such as cars that come down assembly lines by the tens of thousands. Yet with the rise of custom manufacturing, where suppliers create small batches of products on demand, the time it takes to set up a process such as welding or machining becomes a major bottleneck. It takes far too long to prep the robot for its job—sometimes months. People must plan the welding sequence, fasten the parts, program the robot, prepare stock material and optimize welding parameters.

Partnering someone like Dawson Haggerty with a manufacturing robot could cut setup time down dramatically. In the past, programmers used special code to tell robots how to move. Now a product's computer-aided design (CAD) file is all that's needed to set up a smart assembly line. Algorithms will translate these designs into the robot's to-do list.

Designing an assembly line is not the only challenge, however. Robots and people have had a hard time working together. Industrial robots move from position to position and essentially insist on reaching their final destination—whether or not a person is in the way. Manufacturers program their robots to do the same task over and over again until the parts run out. If a rigid object makes a move impossible, industrial robots go into an error state and basically power down. This condition is better than the alternative of going through someone's head, but neither is helpful. Consider how much would get done if coworkers just froze when they got too close to one another.

Next-generation industrial robots will be intrinsically safe around humans. If a robot accidentally hits a human, the blow should not be fatal or even dangerous. Machines will have awareness of where the people are in their workspace, and they should be able to communicate with their human counterparts using voices, gestures, "facial" expressions, text and graphics.

Robot makers are already building machines to meet modern manufacturing's workforce needs. Spitfire is based on a robot made by Zurich-based ABB, augmented with special features designed and built at Carnegie Mellon. ABB also offers Frida, a Two-armed robot designed to operate safely around people. Meanwhile, Boston-based Rethink Robotics, established by iRobot co-founder Rodney Brooks, has developed Baxter, which has two arms as well as an array of sensors to make programming easier than it was for previous generations of robots.

(From *Scientific American*, 2013.5. pp. 40-41. by Bourne Dicken)

☞New Words

1. ubiquitous [juːˈbɪkwɪtəs] *adj.* being present everywhere at once 普遍存在的；无所不在的
2. cohort [ˈkəuhɔːt] *n.* a company of companions or supporters 一群
3. assembly-line *n.* the mechanical system in a factory whereby an article is conveyed through sites at which successive operations are performed on it 装配线；流水作业线
4. laser [ˈleɪzə(r)] *n.* an acronym for light amplification by stimulated emission of radiation; an optical device that produces an intense monochromatic beam of coherent light 激光
5. raw material *n.* the basic material from which a product is made 原料，原材料
6. enormous [ɪˈnɔːməs] *adj.* extraordinarily large in size or extent or amount or power or degree 庞大的，巨大的
7. autonomous [ɔːˈtɔnəməs] *adj.* (of a country or region) having self-government, at least to a significant degree 自治的；独立自主的
8. thermostat [ˈθɜːməstæt] *n.* a regulator for automatically regulating temperature by starting or stopping the supply of heat 恒温器；自动调温器
9. bottleneck [ˈbɒtlnek] *n.* the narrow part of a bottle near the top 瓶颈；障碍物
10. deploy [dɪˈplɔɪ] *v.* to distribute systematically or strategically 配置；展开
11. sequence [ˈsiːkwəns] *n.* a particular order in which related things follow each other 顺序
12. parameter [pəˈræmɪtə(r)] *n.* any factor that defines a system and determines (or limits) its performance 参数
13. algorithm [ˈælgərɪðəm] *n.* a process or set of rules to be followed in calculations or other problem-solving operations, especially by a computer 算法，计算程序
14. rigid [ˈrɪdʒɪd] *adj.* incapable of or resistant to bending 坚硬的
15. intrinsically [inˈtrɪnsɪkli] *adv.* with respect to its inherent nature 本质地
16. tremendous [trəˈmendəs] *adj.* extraordinarily large in size or extent or amount or power or degree 极大的，巨大的

☞Phrases and Expressions

1. burst into: to break open or apart suddenly and violently 闯入
2. be equipped with: to provide with (something) usually for a specific purpose 装配，配备，装备

3. do away with: to terminate, end, or take out 废除，去掉，终止

☞Notes

1. The Robotics Institute (RI) is a division of the School of Computer Science at Carnegie Mellon University in Pittsburgh, Pennsylvania, United States. A June 2014 article in Robotics Business Review magazine calls it "the world's best robotics research facility" and a "pacesetter in robotics research and education."卡耐基梅隆大学机器人研究所是卡耐基梅隆大学计算机学院的分支，位于美国宾夕法尼亚州匹兹堡。2014 年 6 月《机器人商业评论》一篇文章声称，该研究所为"全球最顶尖的机器人研究所"，是"该行业的研究与教学先驱"。
2. Ford Motor Company is an American multinational automaker headquartered in Dearborn, Michigan, a suburb of Detroit. It was founded by Henry Ford and incorporated on June 16, 1903. 福特汽车公司是一家美国大型汽车生产商，由亨利福特于 1903 年 6 月 16 日成立，总部位于美国底特律郊外的密歇根州迪尔伯恩市。
3. General Motors Company, commonly referred to as General Motors (GM), is an American multinational corporation headquartered in Detroit that designs, manufactures, markets, and distributes vehicles and vehicle parts, and sells financial services. 通用汽车公司，即"通用"公司，是一家美国跨国公司，总部位于底特律，业务涉及汽车设计、生产、营销、零部件分销及提供金融服务。

☞Exercises

Vocabulary

Fill in the blanks in each sentence by selecting the most suitable words.

1. In winter, drivers have trouble stopping their cars from ____ on icy roads.
   A. skating B. skidding C. sliding D. slipping
2. This project would ____ a huge increase in defense spending.
   A. result B. assure C. entail D. accomplish
3. The chances of a repetition of these unfortunate events are ____ indeed.
   A. distant B. slim C. unlikely D. narrow
4. We should make a clear ____ between "competent" and "proficient" for the purposes of our discussion.
   A. separation B. division C. distinction D. difference
5. In the present economic ____, we can make even greater progress than previously.
   A. air B. mood C. area D. climate

6. *Rite of Passage* is a good novel by any standards; _____, it should rank high on any list of science fiction.

   A. consistently B. consequently C. invariably D. fortunately

7. The diversity of tropical plants in the region represents a seemingly _____ source of raw materials, of which only a few have been utilized.

   A. exploited B. controversial C. inexhaustible D. remarkable

8. While he was in Beijing, he spent all his time _____ some important museums and buildings.

   A. visiting B. traveling C. watching D. touring

9. You must let me have the annual report without _____ by ten o'clock tomorrow morning.

   A. failure B. hesitation C. trouble D. fail

10. As the director can't come to the reception, I'm representing the company _____.

   A. on his account B. on his behalf C. for his part D. in his interest

Unit 6 Mechanical Engineering

✧ Text A

Mechanical Joint

A mechanical joint is a section of a machine which is used to connect one or more mechanical part to another. Mechanical joints may be temporary or permanent; most types are designed to be disassembled. Most mechanical joints are designed to allow relative movement of these mechanical parts of the machine in one degree of freedom (insert LINK), and restrict movement in one or more others. Mechanical joints are much cheaper and are usually bought ready assembled.

Knuckle joint

A knuckle joint is a mechanical joint used to connect two rods that are under a tensile load when there is a requirement of small amount of flexibility, or angular moment is necessary. There is always an axial or linear line of action of load. The knuckle joint assembly consists of the following major components: 1. Single eye. 2. Double eye or fork. 3. Knuckle pin.

At one end of the rod, the single eye is formed and a double eye is formed at the other end of the rod. Both the single and double eye are connected by a pin inserted through the eye. The pin has a head at one end and at another end there is a taper pin or split pin. For the gripping purpose, the ends of the rod are of octagonal forms. Now, when the two eyes are pulled apart, the pin holds them together. The solid rod portion of the joint in this case is much stronger than the portion through which the pin passes.

Turnbuckle

The buckle or a coupler is a mechanical joint used to connect two members which are subjected to tensile loading which requires slight adjustment of length or tension under loaded conditions. It consists of a central hexagonal nut called coupler and a tie rod having right-hand and left-hand threads. A coupler of hexagonal shape is to facilitate the turning of it with a spanner or sometimes a hole is provided in the nut so that tummy bar can be inserted for rotating it. As the coupler rotates, the tie rod is either pulled together or pushed apart depending upon the direction of the rotation of the coupler. Normally, the tie rods are made of steel while coupler is made of steel or C.I.

Turnbuckles with various sizes are popularly used in construction. They combine strength and durability, yet are simple to set up and adjust. Very small turnbuckles (as light as 10 grams) might be used to support a fence in a garden. On the other hand, extremely large turnbuckles (as heavy as several thousand kilograms) are widely used as supporting high-rise buildings or in structures such as bridges.

Pin joint

A revolute joint (also called pin joint or hinge joint) is a one-degree-of-freedom kinematic pair used in mechanisms. Revolute joints provide a single-axis rotation function used in many places such as door hinges, folding mechanisms, and other uni-axial rotation devices.

Cotter joint

This is used to connect rigidly two rods that transmit motion in the axial direction, without rotation. These joints may be subjected to tensile or compressive forces along the axes of the rods. The very famous example is the joining of the piston rod's extension with the connecting rod in the cross-head assembly.

Bolted joint

Bolted joints are one of the most common elements in construction and machine design. They consist of fasteners that capture and join other parts, and are secured with the mating of screw threads. There are two main types of bolted joint designs: tension joints and shear joints.

In the tension joint, the bolt and clamped components of the joint are designed to transfer an applied tension load through the joint by way of the clamped components by the design of a proper balance of joint and bolt stiffness. The joint should be designed such that the clamp load is never overcome by the external tension forces acting to separate the joint. If the external tension forces overcome the clamp load (bolt preload) the clamped joint components will separate, allowing relative motion of the components.

The second type of bolted joint transfers the applied load in shear of the bolt shank and relies on the shear strength of the bolt. Tension loads on such a joint are only incidental. A preload is still applied but consideration of joint flexibility is not as critical as in the case where loads are transmitted through the joint in tension. Other such shear joints do not employ a preload on the bolt as they are designed to allow rotation of the joint about the bolt, but use other methods of maintaining bolt/joint integrity. Joints that allow rotation include clevis linkages, and rely on a locking mechanism (like lock washers, thread adhesives, and locknuts).

Screw joint

A screw joint is a one-degree-of-freedom kinematic pair used in mechanisms. Screw joints provide single-axis translation by utilizing the threads of the threaded rod to provide such translation. This type of joint is used primarily on most types of linear actuators and certain types of Cartesian robots.

A screw joint is sometimes considered as a separate type but it is actually a variation of bolted joint. The difference is that a screw is used rather than bolt, thus requiring an internal thread in one of the jointed parts. This can save space; however, continuous reuse of the thread would probably damage the coils, making the whole part unsuitable.

(From *Investigations on Microstructure and Mechanical Properties of the Cu/Pb-free Solder Joint Interfaces*, 2016, 6. pp. 78-80. by Robert Hudson)

☞New Words

1. joint [dʒɔɪnt] *n.* (anatomy) the point of connection between two bones or elements of a skeleton, junction by which parts or objects are joined together 关节；接合处
2. section [ˈsekʃn] *n.* a self-contained part of a larger composition, one of the portions into which something is regarded as divided and which together constitute a whole 截面；部分
3. temporary [ˈtemprəri] *adj.* not permanent; not lasting 暂时的；临时的
4. permanent [ˈpɜːmənənt] *adj.* continuing or enduring without marked change in status or condition or place 永久的；不变的
5. disassemble [ˌdɪsəˈsembl] *v.* to take (sth.) to pieces 拆开
6. restrict [rɪˈstrɪkt] *v.* to put a limit on; to keep under control 限定；约束
7. knuckle [ˈnʌkl] *n.* a joint of a finger when the fist is closed 关节；指关节
8. rod [rɒd] *n.* a thin straight bar, especially of wood or metal 棒；枝条
9. tensile [ˈtensaɪl] *adj.* of or relating to tension, capable of being shaped or bent or drawn out 拉力的；可伸长的；可拉长的
10. flexibility [ˌfleksəˈbɪləti] *n.* the property of being flexible; easily bent or shaped 灵活性；弹性；适应性
11. angular [ˈæŋgjələ(r)] *adj.* having angles or an angular shape 有角的；生硬的
12. linear [ˈlɪniə(r)] *adj.* of or in or along or relating to a line; involving a single dimension 线的，线型的；直线的，线状的
13. split [splɪt] *n.* a long crack or hole made when sth. tears 裂缝；裂口
14. pin [pɪn] *n.* a thin piece of metal with a sharp point at one end and a round head at the other（关节）销
15. gripping[ˈgrɪpɪŋ] *adj.* capable of arousing and holding the attention 扣人心弦的；引人注意的
16. octagonal [ɒkˈtægənl] *adj.* of or relating to or shaped like an octagon 八角形的；八边形的

17. dimension [daɪˈmenʃn] *n.* a measurement in space, for example the height, width or length of sth. 维（构成空间的因素）
18. diameter [daɪˈæmɪtə(r)] *n.* the length of a straight line passing through the center of a circle and connecting two points on the circumference 直径；对径
19. shear [ʃɪə(r)] *n.* a strain produced by pressure in the structure of a substance, when its layers are laterally shifted in relation to each other 切变
20. crushing [ˈkrʌʃɪŋ] *adj.* physically or spiritually devastating, often used in combination 压倒的；决定性的
21. turnbuckle [ˈtɜːnbʌkl] *n.* an oblong metal coupling with a swivel at one end and an internal thread at the other into which a threaded rod can be screwed in order to form a unit that can be adjusted for length or tension 套筒螺母；螺丝扣
22. buckle [ˈbʌkl] *n.* a fastener that fastens together two ends of a belt or strap; often has loose prong 皮带扣；带扣
23. hexagonal [heksˈægənl] *adj.* having six sides or divided into hexagons 六边的；六角形的
24. facilitate [fəˈsɪlɪteɪt] *vt.* to make an action or a progress possible or easier 促进；帮助；使容易
25. spanner [ˈspænə(r)] *n.* a hand tool that is used to hold or twist a nut or bolt 扳手；螺丝扳手
26. wagon [ˈwægən] *n.* any of various kinds of wheeled vehicles drawn by a horse or tractor 货车；四轮马车
27. durability [ˌdjʊərəˈbɪləti] *n.* permanence by virtue of the power to resist stress or force 耐久性；坚固；耐用年限
28. hinge [hɪndʒ] *n.* a joint that holds two parts together so that one can swing relative to the other 铰链，合页；关键，转折点
29. capture [ˈkæptʃə(r)] *v.* to take into one's possession or control by force 控制；捕获
30. clamp [klæmpt] 1) *v.* to fasten things firmly together 夹紧 2) *n.* a brace, band, or clasp for strengthening or holding things together 夹具
31. external [ɪkˈstɜːnl] *adj.* happening or arising or located outside or beyond some limits or especially surface 外部的；表面的
32. preload [ˌpriːˈləʊd] *n.* a thing loaded or applied as a load beforehand 预先装载物
33. actuator [ˈæktʃʊeɪtə(r)] *n.* a mechanism that puts something into automatic action 执行机构；激励者；促动器

☞Phrases and Expressions

1. tensile load: a large quantity of something that is carried by a vehicle, person etc.拉伸负荷

2. in this case: in the circumstance 既然这样，在这种情况下
3. be subjected to: to receive, suffer 受到……；经受……
4. set up: to place or erect something in position 建立；装配
5. …be secured with: to use something to fasten something 用……固定

☞Notes

1. one degree of freedom（单自由度，一个自由度）: Degrees of freedom are the number of values in a study that have the freedom to vary. They are commonly discussed in relation to various forms of hypothesis testing in statistics, such as a chi-square. It is essential to calculate degrees of freedom when we try to understand the importance of a chi-square statistic and the validity of the null hypothesis.自由度是研究中可以自由变化的数值。它们通常与统计学中的各种形式的假设试验（如卡方）的关系进行讨论。计算自由度是理解卡方统计的重要性和零假设的有效性的十分重要的方法。
2. Cartesian robots: Cartesian coordinate robots (also called linear robots) are industrial robots whose three principal axis of control are linear (i.e. they move in a straight line rather than rotate) and are at right angles to each other. The three sliding joints correspond to moving the wrist up-down, in-out, back-forth. Among other advantages, this mechanical arrangement simplifies the robot control arm solution.笛卡儿坐标机器人（也称为线性机器人）是一种工业机器人，其三个主要的控制轴是线性的（即它们以直线而不是旋转的方式移动）并且彼此成直角。三个滑动关节用于控制手腕上下，内外，后退移动。它还有其他优点，其一就是这种机械布置简化了机器人控制臂的解决方案。

☞Exercises

Ⅰ. Reading Comprehension

Answer the following questions according to the text.

1. What are the major components of the knuckle joint?
2. What are knuckle joint's applications?
3. How many uses does turnbuckle have? What are they?
4. What are the types of bolted joint design?
5. Please list the advantages of the cotter joint.

Ⅱ. Vocabulary

A. Replace the underlined words with the correct form of the words and expressions from the word bank.

A. adapt	B. apart from	C. although	D. doubtful	E. energetic	F. expectation
G. greedy	H. investigate	I. the norm	J. plentiful	K. a little	L. and so on
M. as well	N. have no right to	O. and name after			

Fast food has become a habit (1) for many university students, even though (2) there is information about the dangers of unhealthy eating which is found everywhere (3). Food scientists in the UK have researched (4) the effect of students' diets on their ability to study. Over a three-month period, students changed (5) their diets. Except for (6) milk products, they ate very little animal fat. As to ideas (7) about what would happen, most students said they felt more active (8) than before, and some said they no longer felt they wanted to eat more than they need (9) at mealtimes. But it is not certain (10) that the majority of students will change their habits as a result of the experiment.

B. Fill in the blanks in each sentence by selecting the most suitable words.

1. The rain was heavy and ____ the land was flooded.
   A. consequently B. continuously C. consistently D. constantly
2. A lot of ants are always invading my kitchen. They are a thorough ____.
   A. nuisance B. trouble C. anxiety D. worry
3. It is our ____ policy that we will achieve unity through peaceful means.
   A. consistent B. continuous C. considerate D. continual
4. When people become unemployed, it is ____ which is often worse than lack of wages.
   A. laziness B. poverty C. idleness D. inability
5. Some people would like to do shopping on Sundays since they expect to pick up wonderful ____ in the market.
   A. batteries B. bargains C. baskets D. barrels
6. There were no tickets ____ for Friday's performance.
   A. preferable B. considerable C. possible D. available
7. American women were ____ the right to vote until 1920 after many years of hard struggle.
   A. ignored B. neglected C refused D. denied

8. Silver is the best conductor of electricity, copper ____ it closely.

A. followed B. following C. to follow D. being followed

9. What you have done is ____ the doctor's orders.

A. attached to B. responsible to C. resistant to D. contrary to

10. Young people are not ____ to stand and look at works of art; they want art they can participate in.

A. conservative B content C. confident D. generous

Ⅲ. Cloze

Choose an appropriate word from the four choices marked A, B, C and D for each blank in the passage.

The term shaft usually refers to a relatively long member of the round cross section that rotates and transmits power. One or more members such as gears, sprockets, pulleys, and cams are usually __1__ to the shaft by means of pins, keys, splints, snap rings, and other devices. These latter members are among the "associated parts" considered in this text, as are couplings and universal joint, which serve to __2__ the shaft to the source of power or load.

A shaft can have a no round cross section, and it need not necessarily rotate. It can be stationary and serve to __3__ a rotating member, such as the short shafts that support the non-driving wheels of an automobile. The shafts supporting __4__ gears can be either rotating or stationary depending on __5__ the gear is attached to the shaft or supported by it through bearings.

It is apparent that shafts can be subjected to various combinations of axial, bending, and __6__ loads, and that these loads may be static or fluctuating. Typically, a rotating shaft transmitting power is __7__ to a constant torque (producing mean torsion stress) together with a completely reversed bending load (producing alternating bending stress).

In addition to satisfying __8__ requirements, shafts must be designed so that deflections are within acceptable limits. Excessive __9__ shafts deflection can hamper gear performance and cause __10__ noises.

1. A. cement B. attached C. connected D. concrete
2. A. endure B. transmit C. serve D. connect
3. A. support B. meet C. satisfy D. strong
4. A. aims B. idler C. terminal D. tomb
5. A. due to B. weather C. whether D. for
6. A. causticity B. torsional C. inertia D. acceleration
7. A. subjected B. connects C. and D. from
8. A. deeply B. strength C. clearly understand D. long time
9. A. identify B. cross C. round D. lateral

10. A. lucrative B. objectionable C. attractable D. pleasant

Ⅳ. Translation

Translate the following text into Chinese.

As we look around us, we see a world full of "things", machines, devices, tools, things that we have designed, built, and used; things made of wood, metals, ceramics, and plastics. We know from experience that some things are better than others; they last longer, cost less, are quieter, or are case to use. Ideally, however, every such item has been designed according to some set of "functional requirements" as perceived by the designers—that is, it has been designed so as to answer the question, "exactly what function should it perform?" In the world of engineering, the major function frequently is to support some type of loading due to weight, inertia, pressure, etc.

Ⅴ. Writing

Write an essay on the following topic with no less than 150 words. Give reasons for your answer and include relevant examples from your own knowledge or experience.

Briefly Introduce Your Understanding of the Current Mechanical Engineering Profession

✧ Text B

Mechanical Resonance

Mechanical resonance is the tendency of a mechanical system to respond at greater amplitude when the frequency of its oscillations matches the systems' natural frequency of vibration (its *resonance frequency* or *resonant frequency*) than it does at other frequencies. It may cause violent swaying motions and even catastrophic failure in improperly constructed structures including bridges, buildings and airplanes. This is a phenomenon known as resonance disaster.

Avoiding resonance disasters is a major concern in every building, tower and bridge construction project. The Taipei 101 building relies on a 660-ton pendulum—a tuned mass damper—to modify the response at resonance. Furthermore, the structure is designed to resonate at a frequency that does not typically occur. Buildings in seismic zones are often constructed to take into account the oscillating frequencies of expected ground motion.

In addition, engineers designing objects having engines must ensure that the mechanical resonant frequencies of the component parts do not match driving vibration frequencies of the motors or other strongly oscillating parts. Many resonant objects have more than one resonance frequency. It will vibrate easily at those frequencies, and less so at other frequencies. Many clocks keep time by mechanical resonance in a balance wheel, pendulum, or quartz crystal.

The natural frequency of a simple mechanical system consisting of a weight suspended by a spring is: Where m is the mass and k is the spring constant. A swing set is a simple example of a resonant system with which most people have practical experience. It is a form of the pendulum. If the system is excited (pushed) with a period between pushes equal to the inverse of the pendulum's natural frequency, the swing will swing higher and higher, but if excited at a different frequency, it will be difficult to move. The resonance frequency of a pendulum, the only frequency at which it will vibrate, is given approximately, for small displacements, by the equation: where g is the acceleration due to gravity (about 9.8 m/s^2 near the surface of Earth), and L is the length from the pivot point to the center of mass. (An elliptic integral yields a description for any displacement). Note that, in this approximation, the frequency does not depend on mass.

Mechanical resonators work by transferring energy repeatedly from kinetic to potential form and back again. In the pendulum, for example, all the energy is stored as gravitational energy (a form of potential energy) when the bob is instantaneously motionless at the top of its swing. This energy is proportional to both the mass of the bob and its height above the lowest point. As the bob descends and picks up speed, its potential energy is gradually converted to kinetic energy (energy of movement), which is proportional to the bob's mass and to the square of its speed. When the bob is at the bottom of its travel, it has maximum kinetic energy and minimum potential energy. The same process then happens in reverse as the bob climbs towards the top of its swing. Some resonant objects have more than one resonance frequency, particularly at harmonics (multiples) of the strongest resonance. It will vibrate easily at those frequencies, and less so at other frequencies. It will pick out its resonance frequency from a complex excitation, such as an impulse or a wideband noise excitation. In effect, it is filtering out all frequencies other than its resonance. In the example above, the swing cannot easily be excited by harmonic frequencies, but can be excited by subharmonics.

Resonance may cause violent swaying motions in improperly constructed structures, such as bridges and buildings. The London Millennium Footbridge (nicknamed the *Wobbly Bridge*) exhibited this problem. A faulty bridge can even be destroyed by its resonance (see Broughton Suspension Bridge and Angers Bridge). Mechanical systems store potential energy in different forms. For example, a spring/mass system stores energy as tension in the spring, which is ultimately stored as the energy of bonds between atoms.

In mechanics and construction, a resonance disaster describes the destruction of a building or a technical mechanism by induced vibrations at a system's resonance frequency, which causes it to oscillate. Periodic excitation optimally transfers to the system the energy of the vibration and stores it there. Because of this repeated storage and additional energy input, the system swings ever more strongly until its load limit is exceeded.

Tacoma Narrows Bridge

The dramatic, rhythmic twisting that resulted in the 1940 collapse of "Galloping Bertie", the original Tacoma Narrows Bridge, is sometimes characterized in physics textbooks as a classic example of resonance. The catastrophic vibrations that destroyed the bridge were due to an oscillation caused by interactions between the bridge and the winds passing through its structure—a phenomenon known as aeroelastic flutter. Robert H. Scanlan, the father of the field of bridge aerodynamics, wrote an article about this.

Various methods of inducing mechanical resonance in a medium exist. Mechanical waves can be generated in a medium by subjecting an electromechanical element to an alternating electric field having a frequency which induces mechanical resonance and is below any electrical resonance frequency. Such devices can apply mechanical energy from an external source to an element to mechanically stress the element or apply mechanical energy produced by the element to an external load.

The United States Patent Office classifies devices that test mechanical resonance under the list of resonance, frequency, and amplitude for measuring and testing. This subclass is itself indented under subclass 570, Vibration. Such devices test an article or mechanism by subjecting it to a vibratory force for determining qualities, characteristics, or conditions thereof, or sensing, studying or making analysis of the vibrations otherwise generated in or existing in the article or mechanism. Devices include the right methods to cause vibrations at a natural mechanical resonance and measure the frequency and amplitude the resonance made. Various devices study the amplitude response over a frequency range. This includes nodal points, wavelengths, and standing wave characteristics measured under predetermined vibration conditions.

(From *Mechanical Engineering*, 1999(3), pp. 263-265. by Scott Reeves)

☞New Words

1. resonance [ˈrezənəns] *n.* an excited state of a stable particle causing a sharp maximum in the probability of absorption of electromagnetic radiation 共振；共鸣；反响
2. amplitude [ˈæmplɪtjuːd] *n.* (physics) the maximum displacement of a periodic wave; the property of copious abundance 振幅；丰富；充足
3. oscillation [ˌɒsɪˈleɪʃn] *n.* the process of oscillating between states; (physics) a regular periodic variation in value about a mean 在多种状态间振荡的过程；（物理）状态平均值间的一种有规律的周期性变化

4. vibration [vaɪˈbreɪʃn] *n.* a continuous shaking movement or feeling 震动；颤动；抖动；（感情的）共鸣
5. sway [sweɪ] *v.* to move back and forth or sideways; to cause to move back and forth 使摇动；使摇摆
6. catastrophic [ˌkætəˈstrɒfɪk] *adj.* extremely harmful; bringing physical or financial ruin 灾难性的，毁灭性的
7. pendulum [ˈpendjələm] *n.* an apparatus consisting of an object mounted so that it swings freely under the influence of gravity 钟摆；摇锤
8. damper [ˈdæmpə(r)] *n.* a movable iron plate that regulates the draft in a stove or chimney or furnace; a device that decreases the amplitude of electronic, mechanical, acoustical, or aerodynamic oscillations 减震器；气闸
9. approximately [əˈprɒksɪmətli] *adv.* (of quantities) imprecise but fairly close to correctness 大约，近似地；近于
10. displacement [dɪsˈpleɪsmənt] *n.* the action of moving something from its place or position 取代；移位
11. acceleration [əkˌseləˈreɪʃn] *n.* an increase in the rate of change; the act of accelerating; increasing the speed; (physics) a rate of increase of velocity 加速，促进；[物] 加速度
12. pivot [ˈpɪvət] *n.* the axis consisting of a short shaft that supports something that turns; the act of turning on (or as if on) a pivot 枢轴；中心点；旋转运动
13. approximation [əˌprɒksɪˈmeɪʃn] *n.* an estimate of a number or an amount that is almost correct, but not exact 近似值；粗略估计
14. kinetic [kɪˈnetɪk] *adj.* relating to or resulting from motion 运动的
15. gravitational [ˌɡrævɪˈteɪʃənl] *adj.* of or relating to or caused by gravitation 重力的，引力的
16. instantaneously [ˌɪnstənˈteɪniəsli] *adv.* without any delay 即刻；突如其来地
17. proportional [prəˈpɔːʃənl] *adj.* increasing or decreasing in size, amount or degree according to changes in sth. else 成比例的；相称的
18. descend [dɪˈsend] *v.* to move downward and lower, but not necessarily all the way 下降；下去
19. harmonic [hɑːˈmɒnɪk] *adj.* of or relating to harmony as distinct from melody and rhythm 和声的；音乐般的
20. impulse [ˈɪmpʌls] *n.* an instinctive motive 冲动；神经冲动；推动力
21. ultimately [ˈʌltɪmətli] *adv.* as the end result of a succession or process 最后；根本；基本上
22. oscillate [ˈɒsɪleɪt] *v.* to keep moving from one position to another and back again 振动；摆动
23. periodic [ˌpɪəriˈɒdɪk] *adj.* happening or recurring at regular intervals 周期的；定期的

24. rhythmic [ˈrɪðmɪk] *adj.* recurring with measured regularity 有节奏的,合拍的
25. catastrophic [ˌkætəˈstrɒfɪk] *adj.* extremely harmful; bringing physical or financial ruin 灾难的；悲惨的
26. electromechanical [ɪˈlektrəʊməˈkænɪkl] *adj.* of or relating to or involving an electrically operated mechanical device 电动机械的，机电的
27. vibratory [vaɪˈbreɪtəri] *adj.* moving very rapidly to and fro or up and down 振动性的；震动的
28. nodal [ˈnəʊdl] *adj.* of or like a node 节的
29. predetermined [ˌpriːdɪˈtɜːmɪnd] *adj.* established or decided in advance 预先决定的

☞Phrases and Expressions

1. mechanical resonance: the additional vibrations and echoes of the original sound 机械共振
2. seismic zones: areas where earthquakes occur frequently 地震带
3. in addition: by way of addition; furthermore 另外，此外
4. take into account: to allow or plan for a certain possibility; to concede the truth or validity of something 考虑；重视
5. balance wheel: a wheel that regulates the rate of movement in a machine; especially a wheel oscillating against the hairspring of a timepiece to regulate its beat 摆轮；平衡轮
6. quartz crystal: a thin plate or small rod of quartz cut along certain lines and ground so that it can produce an electric signal at a constant frequency; used in crystal oscillators 石英晶体
7. consist of: to be constitutive of 由……组成；由……构成；包括
8. equal to: having the requisite qualities for 等于；胜任
9. gravitational energy: the potential energy associated with the gravitational field 重力场位能

☞Notes

1. The London Millennium Footbridge is a steel suspension bridge for pedestrians crossing the River Thames in London, linking bankside with the City of London. It is owned and maintained by Bridge House Estates, a charitable trust overseen by the City of London Corporation. Construction began in 1998, and it initially opened in June 2000. 伦敦千禧桥，是一座钢制吊桥，可供行人在伦敦的泰晤士河上通行，同时将河岸与伦敦市区连接起来。其所有权和维护权归伦敦金融城公司监管的慈善信托公司“桥屋”所有。该桥始建于 1998 年，于 2000 年 6 月首次开放。

2. Tacoma Narrows Bridge: a suspension bridge at Tacoma in Washington 塔科马海峡大桥：位于华盛顿塔科马的悬索桥
3. The United States Patent Office 美国专利局

☞Exercises

Vocabulary

Replace the underlined words with the correct form of the words and expressions from the word bank. Change the form when necessary.

put down	stick to	cut down on	lead to	get one's hands on	put in
make the most of	put away	time after time	count for	much more	
give away	had in mind	most of the time	go off	sort out	

1. The doctor suggested that both of them spend 20 minutes a day doing Tai Chi.
2. Tom studied hard. He wanted to make the best use of his time in college to learn as much as he could.
3. When taking notes, you don't have to write down everything the teacher says.
4. Active reading, according to education experts, will result in a better understanding of what being read.
5. During those years in the countryside, she was reading everything she could find.
6. It's going to rain: we'd better put our picnic things in proper places and go indoors.
7. In spite of the new difficulties, we've decided not to change our plan.
8. I have told you again and again not to exercise immediately after lunch or supper.
9. Taking his doctor's advice, my uncle has smoked less as the first step.
10. Knowing how to make most of your abilities is much more important than working hard.

Unit 7 Economics and Management

✧ Text A

India's Economy: The Missing Middle Class

After China, where next? Over the past two decades, the world's most populous country has become the market indispensable to just about every global company seeking growth. As its economy slows, businesses are looking for the next set of consumers to keep the tills ringing.

To many, India feels like the heir apparent. Its population will soon overtake its Asian rival's. It occasionally grows at the kind of pace that propelled China to the status of an economic superpower. And its middle class is thought by many to be in the early stages of the journey to prosperity that created hundreds of millions of Chinese consumers. Exuberant management consultants speak of a 300m-400m horde of potential frappuccino-sippers, Fiesta-drivers and globe-trotters. Rare is the chief executive who, upon visiting India, does not proclaim it as central to his or her plans. Some of that may be a diplomatic dose of flattery; much of it, from firms such as IKEA, SoftBank, Amazon and Starbucks, is sincerely meant.

Hold your elephants. The Indian middle class conjured up by the marketers and consultants scarcely exists. Firms peddling anything much beyond soap, matches and phone-credit are targeting a minuscule slice of the population. The top 1% of Indian adults, a rich enclave of 8m inhabitants making at least $20,000 a year, equates to roughly Hong Kong of China in terms of population and average income. The next 9% is akin to central Europe, in the middle of the global wealth pack. The next 40% of India's population neatly mirrors its combined South Asian poor neighbours, Bangladesh and Pakistan. The remaining half-billion or so are on a par with the most destitute bits of Africa. To be sure, global companies take the markets of central Europe seriously. Plenty of fortunes have been made there.

Centre parting

Worse, the chances of India developing a middle class to match the Middle Kingdom's

are being throttled by growing inequality. The top 1% of earners pocketed nearly a third of all the extra income generated by economic growth between 1980 and 2014, according to new research from economists including Thomas Piketty. The well-off are ten times richer now than in 1980; those at the median have not even doubled their income. India has done a good job at getting those earning below $2 a day (at purchasing-power parity) to $3, but it has not matched other countries' records in getting those on $3 a day to earning $5, those at $5 a day to $10, and so on. Middle earners in countries at India's stage of development usually take more of the gains from growth. Eight in ten Indians cite inequality as a big problem, on a par with corruption.

The reasons for this failure are not mysterious. Decades of statist intervention meant that when a measure of liberalization came in the early 1990s, only a few were able to benefit. The workforce is woefully unproductive—no surprise given the state of India's education system, which churns out millions of adults equipped only for menial work. Its graduates go on to toil in small or micro-enterprises, operating informally; these "employ" 93% of all Indians. The great swell of middle-class jobs that China created as it became the workshop to the world is not to be found in India, because turning small businesses into productive large ones is made nigh on impossible by bureaucracy in India. The fact that barely a quarter of women work—a share that has seen a precipitous decline in the past decade—only makes matters worse.

Good policy can do an enormous amount to improve prospects. However, hope should be tempered by realism. India is blessed with a deeply entrenched democratic system, but that is no shield against poor decisions. The sudden and brutal "demonetization" of the economy in 2016 was meant to target fat cats, but ended up hurting everybody. And the path to prosperity walked by China, where manufacturing produced the jobs that pushed up incomes, is narrowing as automation limits opportunities for factory work.

All of which means that companies need to deal with the India that exists today rather than the one they wish to emerge. A strategy of waiting for Indians to develop a taste for products that the global middle class indulges in—cars as income per head crosses one threshold, foreign holidays when it crosses the next—may lead to decades of frustration. Only 3% of Indians have ever been on an aeroplane; only one in 45 owns a car or lorry. If nearly 300m Indians count as "middle class", as HSBC has proclaimed, some of them make around $3 a day.

Big market, smaller opportunities

Companies would do better to "Indianise" their business by, for example, peddling wares using regional languages preferred by hundreds of millions of Indians. Pricing matters. Services proffered at the same price in India as Indiana will appeal to mere millions, not a billion. Even for someone in the top 10% of Indian earners, an annual Netflix subscription can cost over a week's income; the equivalent in America would be around $3,000. Apple ads may plaster Mumbai, Delhi and Bangalore, but for only one in ten Indians would the latest iPhone represent less than half a year's salary. The biggest consumer hits in India have been goods and services that offer stonking value: scooters and mobile telephony have grown fast, but only after prices tumbled.

The sharpest businesses work out which "enablers" will allow Indians to gain access to new goods. Electrification drives demand for fridges. Cheap mobile data (India is in the midst of a data-price war that has hugely benefited consumers) are a boon to streaming services. Logistics networks put together by e-commerce giants are for the first time making it possible for a consumer in a third-tier city to buy global fashion brands. A surge in consumer financing has put desirable baubles within reach of more Indians.

Insofar as it is the job of politicians to create a consumer class, successive Indian governments have largely failed. Businesses hoping the Indian middle class will provide their next spurt of growth should be under no illusion. Companies will have to work very hard to turn potential into profits.

(From *The Economist*, 2018, 1. pp.16-17. by John Micklethwait & Daniel Franklin)

☞New Words

1. till [tɪl] *n.* a strongbox for holding cash 放钱的抽屉
2. heir [eə(r)] *n.* a person who inherits some title or office 继承者
3. exuberant [ɪɡˈzjuːbərənt] *adj.* full of energy, excitement and happiness 精力充沛的；热情洋溢的；兴高采烈的
4. consultant [kənˈsʌltənt] *n.* an expert who gives advice 顾问
5. diplomatic [ˌdɪpləˈmætɪk] *adj.* skilled in dealing with sensitive matters or people 老练的
6. minuscule [ˈmɪnəskjuːl] *adj.* very small 极小的
7. enclave [ˈenkleɪv] *n.* an enclosed territory that is culturally distinct from the foreign territory that surrounds it 被包围的领土
8. neatly [ˈniːtli] *adv.* with neatness 整洁地；熟练地
9. mirror [ˈmɪrə(r)] *vt.* to reflect as if in a mirror 反射；反映
10. throttle [ˈθrɒtl] *vt.* to place limits on (extent or access)压制
11. intervention [ˌɪntəˈvenʃn] *n.* the act of intervening (as to mediate a dispute) 介入；调停
12. liberalization [ˌlɪbrəlaɪˈzeɪʃn] *n.* the act of making less strict 自由化
13. woefully [ˈwəʊfəli] *adv.* in an unfortunate or deplorable manner 不幸地
14. menial [ˈmiːniəl] *adj.* not requiring much skill and lacking prestige 不需技巧的；卑微的
15. nigh [naɪ] *adj.* not far distant in time or space or degree or circumstances 在附近的
16. bureaucracy [bjʊəˈrɒkrəsi] *n.* the system of official rules and ways of doing things that a

government or an organization has, especially when these seem to be too complicated 官僚主义；官僚作风

17. precipitous [prɪˈsɪpɪtəs] *adj.* extremely steep; done with very great haste and without due deliberation 险峻的；急躁的
18. temper [ˈtempə(r)] *vt.* to make more temperate, acceptable, or suitable by adding something else; to moderate 使调和
19. entrenched [ɪnˈtrentʃt] *adj.* established firmly and securely 根深蒂固的；确立的；不容易改的
20. demonetization [diːmʌnɪtaɪˈzeɪʃn] *n.* the withdrawal of a coin, note, or precious metal from use as legal tender 废止流用
21. proffer [ˈprɒfə(r)] *vt.* to present for acceptance or rejection 提供；提出；奉献
22. plaster [ˈplɑːstə(r)] *vt.* to cover conspicuously or thickly, as by pasting something on 张贴
23. stonk [stɒŋk] *vt.* to bombard (soldiers, buildings, etc.) with artillery 猛烈炮轰
24. tumble [ˈtʌmbl] *vi.* to fall down, as if collapsing 摔倒
25. electrification [ɪˌlektrɪfɪˈkeɪʃn] *n.* the connecting of the places to a supply of electricity 电气化
26. logistics [ləˈdʒɪstɪks] *n.* the business of transporting and delivering goods 物流
27. bauble [ˈbɔːbl] *n.* a piece of jewellery that is cheap and has little artistic value 廉价首饰
28. spurt [spɜːt] *vi.* to gush forth in a sudden stream or jet 迸发

☞Phrases and Expressions

1. indispensable: an essential action, condition, or ingredient 必要条件；要素
2. keep the tills ringing: to keep making money 有利可图
3. hold your elephants: to be patient 耐心一点
4. on a par with: equal or similar to someone or something 与……同等
5. churn out: to produce something at a fast rate 大量炮制

☞Notes

1. IKEA: IKEA is a Swedish-founded multinational group that designs and sells ready-to-assemble furniture, kitchen appliances and home accessories, among other useful goods and occasionally home services. 宜家家居是成立于瑞典的一家跨国集团公司，设计和销售即装即用的家具、厨房用品、家居饰品以及其他一些有用的物品，有时提供一些家居服务。
2. SoftBank: The company is known for its leadership by founder Masayoshi Son. It was

founded in 1981 in Japan. It now owns operations in broadband, fixed-line telecommunications, e-commerce, Internet, technology services, finance, media and marketing, semiconductor design, and other businesses. 软件银行集团因创建者孙正义而闻名，它于 1981 年在日本被创立。目前拥有宽带业务、固定电话、电子商务、互联网、技术服务、金融、媒体与营销、半导体设计以及其他业务。

3. HSBC: HSBC Holdings plc is a British multinational banking and financial services holding company. It is the 7th largest bank in the world, and the largest in Europe. 汇丰控股有限公司是一家英国跨国银行和金融服务控股公司。它是世界上第七大银行，也是欧洲最大的银行。

☞Exercises

Ⅰ. Reading Comprehension

Answer the following questions according to the text.

1. What do the exuberant management consultants think will happen?
2. What is the income situation of the Indian people?
3. What did the new research from economists including Thomas Piketty report?
4. What companies would do to "Indianise" their business?
5. What is the biggest consumer hits in India according to the article?

Ⅱ. Vocabulary

A. Replace the underlined words or expressions in the following sentences with one of the best choices from the word bank.

A. entrenched	B. precipitous	C. intervened	D. proffered	E. stonk
F. bureaucracy	G. menial	H. electrification	I. demonetization	J. spurts
K. on a par with	L. churn out	M. sine qua non	N. logistics	O. nigh on

1. The destiny of the era was often <u>as same as</u> the outstanding personages' destiny in that time.
2. The Industrial Age was largely about making those jobs as <u>worthless</u> and unskilled as possible.
3. Perry appears almost desperate to win back conservative voters and reverse his <u>rapid</u> drop.
4. Mental health counseling may prove to be the <u>essence</u> of successfully re-integrating child soldiers into civilian life.
5. The President <u>mediated</u> personally in the crisis.
6. Normally 6, 000 workers <u>produce</u> 240, 000 cars a year, a twentieth of Honda's global output.
7. The army has not yet <u>given</u> an explanation of how and why the accident happened.

8. When the washing machine spouts out water at least we can mop it up.
9. The system was rooted in the early 1980s, but it ran into problems of obsolescence.
10. They've lived in that house for nearly 30 years.

B. Fill in the blanks in each sentence by selecting the most suitable words.

1. The skills and _____ of getting such a big show on the road pose enormous practical problems.
   A. logic B. logical C. logistics D. logician
2. A series of _____ performances in friendliness would suggest the team is not improving.
   A. wonderful B. splendid C. brilliant D. woeful
3. Thick golden curls _____ down over her shoulders.
   A. tunnel B. tumble C. tumult D. tumour
4. The military procurement _____ apparently cannot come up with a set-top decoding box quickly enough.
   A. bureaucratic B. bureaucrat C. bureaucracy D. bureaucratism
5. That little fact rattled him and the rest of his practice session was _____.
   A. abyssal B. abysmal C. abnormal D. absolute
6. The king blamed _____ and taxes for high oil prices and suggested setting up a program of 1 billion U.S. dollars to solve the oil crisis.
   A. peculiarity B. speculation C. spectacle D. prospect
7. A contract is the only document between the parties to which they may _____ for clarification of mutual responsibilities.
   A. refer B. offer C. prefer D. differ
8. He rose and _____ a silver box full of cigarettes.
   A. referred B. proffered C. preferred D. differed
9. He then requires man to work hard, fulfill his duties and meet his _____.
   A. obligations B. resignation C. allowances D. supervision
10. We've worked out a method by which our production can be raised on a large _____.
    A. quantity B. scale C. quality D. proportion

Ⅲ. Cloze

Choose an appropriate word from the four choices marked A, B, C and D for each blank in the passage.

Shopping habits in the United States have changed greatly in the last quarter of the 20th century. __1__ in the 1900s most American towns and cities had a Main Street. Main Street was always in the heart of a town. This street was lined on both sides with many __2__ businesses. Here, shoppers walked into stores to look at all sorts of merchandise: clothing, furniture, hardware, groceries. In addition, some shops offered __3__.

These shops included drugstores, restaurants, shoe-repair stores, and barber or hairdressing shops. But in the 1950s, a change began to __4__. Too many automobiles had crowded into Main Street while too few parking places were __5__ shoppers. Because the streets were crowded, merchants began to look with interest at the open spaces outside the city limits. Open space is what their car-driving customers needed. And open space is what they got __6__ the first shopping centre was built. Shopping centres, or rather malls, started as a collection of small new stores __7__ crowded city centres. Attracted by hundreds of free parking spaces, customers were drawn away from __8__ areas to outlying malls. And the growing popularity of shopping centres led __9__ to the building of bigger and better-stocked stores. By the late 1970s, many shopping malls had almost developed into small cities themselves. In addition to providing the __10__ of one-stop shopping, malls were transformed into landscaped parks, with benches, fountains, and outdoor entertainment.

1. A. As early as	B. Early	C. Early as	D. Earlier
2. A. varied	B. various	C. sorted	D. mixed up
3. A. medical care	B. food	C. cosmetics	D. services
4. A. be taking place	B. take place	C. be taken place	D. have taken place
5. A. available for	B. available to	C. used by	D. ready for
6. A. when	B. while	C. since	D. then
7. A. out of	B. away from	C. next to	D. near
8. A. inner	B. central	C. shopping	D. downtown
9. A. on	B. in turn	C. by turns	D. further
10. A. cheapness	B. readiness	C. convenience	D. handiness

Ⅳ. Translation

Translate the following text into Chinese.

To many, India feels like the heir apparent. Its population will soon overtake its Asian rival's. It occasionally grows at the kind of pace that propelled China to the status of an economic superpower. And its middle class is thought by many to be in the early stages of the journey to prosperity that created hundreds of millions of Chinese consumers. Exuberant management consultants speak of a 300m-400m horde of potential frappuccino-sippers, Fiesta-drivers and globe-trotters. Rare is the chief executive who, upon visiting India, does not proclaim it as central to his or her plans. Some of that may be a diplomatic dose of flattery; much of it, from firms such as IKEA, SoftBank, Amazon and Starbucks, is sincerely meant.

Ⅴ.Writing

Write an essay on the following topic with no less than 150 words. Give reasons for

your answer and include any relevant examples from your own knowledge or experience.

Apart from making money, businesses should also have social responsibilities.
Do you agree or disagree?

✧ Text B

User-rating Systems Are Cut-rate Substitutes for a Skillful Manager

It often arrives as you stroll from the kerb to your front door. An e-mail with a question: how many stars do you want to give your Uber driver? Rating systems like the ride-hailing firm's are essential infrastructure in the world of digital commerce. Just about anything you might seek to buy online comes with a crowd-sourced rating, from a subscription to this newspaper to a broken iPhone on eBay to, increasingly, people providing services. But people are not objects. As ratings are applied to workers it is worth considering the consequences—for the rater and the rated.

User-rating systems were developed in the 1990s. The web held promise as a grand bazaar, where anyone could buy from or sell to anyone else. But e-commerce platforms had to create trust. Buyers and sellers needed to believe that payment would be forthcoming, and that the product would be as described. E-tailers like Amazon and eBay adopted reputation systems, in which sellers and buyers gave feedback about transactions. Reputation scores appended to products, vendors and buyers gave users confidence that they were not about to be scammed.

Such systems then spread to labour markets. Workers for gig-economy firms like Uber and Upwork come with user-provided ratings. Conventional employers are jumping on the bandwagon. A phone call to your bank, or the delivery of a meal ordered online, is now likely to be followed by a notification prompting you to rate the person who has just served you.

Superficially, such ratings also seem intended to build trust. For users of Uber, say, who will be picked up by drivers they do not know, ratings look like a way to reassure them that their ride will not end in the abduction. Yet if that was once necessary, it is no longer. Uber is a global firm worth tens of billions of dollars and with millions of repeat customers. Its customers know by now that the app records drivers' identities and tracks their route. It is Uber's brand that creates trust; for most riders, waiting for a driver with a rating of 4.8 rather than 4.5 is not worth the trouble.

Rather, ratings increasingly function to make management cheaper by shifting the burden of monitoring workers to users. Though Uber regards its drivers as independent contractors, in many ways they resemble employees. The firm seeks to provide users with a reasonably uniform experience from ride to ride. And because drivers are randomly assigned to customers, it is the platform that cares whether rides lead to repeat business and which therefore bears the cost of poor behaviour by drivers. Ordinarily, a firm in such a position would need to invest heavily in monitoring its workers—hiring staff to carry out quality assurance by taking Uber rides incognito, for instance. A rating system, however, reduces the need for monitoring by aligning the firm's interests with those of workers. (Drivers with low ratings risk having their profile deactivated.)

Outsourcing management like this appeals to cost-conscious firms of all sorts; hence the proliferation of technological nudges to rate one service worker or another. To work as intended, however, ratings must provide an accurate indication of how well workers conform to the behaviour that firms desire. Frequently, they do not. Raters may have no incentive to do their job well. They may ignore the prompt to rate a worker, or automatically assign the highest score. They may adhere to social norms that discourage leaving a poor rating, just as diners often leave the standard tip, however unexceptional the service. Uber's customers often award drivers five stars rather than feel bad about themselves for damaging a stranger's work prospects. And even when users are accurate, their ratings may reflect factors beyond a service provider's control, such as unexpected traffic. Systems that allow users to leave more detailed feedback (as Uber's has begun to) could address this, but at the cost of soaking up more time, which could mean fewer reviews.

When the quality of a match between a worker and a task is particularly important, the problem of sorting the signal from the noise in rating systems grows. Skilled managers can tell when a worker struggling in one role might thrive in another; rating systems can capture only expressions of customer dissatisfaction. Such difficulties also affect gig-economy platforms. Poor ratings on a job-placement site could reflect an inappropriate pairing between a worker with one set of skills and a firm that needs another, rather than the worker's failure of effort or ability.

Platforms can reduce the potential for such errors by including more information about tasks and the workers who might tackle them. Yet they may discover to their chagrin that more information also provides users with more opportunities to discriminate. An analysis of Upwork, for example, found that employers of Indian descent disproportionately sought Indian nationals for their tasks. True, this particular sort of information could be concealed—and conventional management permits plenty of discrimination. But firms typically have a legal obligation not to discriminate, and to train managers accordingly.

Management is underappreciated as a contributor to success. Recent work by Nicholas Bloom, John Van Reenen and Erik Brynjolfsson suggests that good management matters more than the adoption of technology for a company's performance. Even so, the use of ratings seems sure to grow. They are, as "Left Outside", a pseudonymous blogger, puts it, a genuinely disruptive technology: cheap enough to be adopted widely even if inferior to established practice. Further advances could improve such systems, as is common with disruptive technology. Artificial-intelligence programs may one day know how much people enjoy a taxi ride better than they do themselves. In the meantime, management risks being left to the wrong sort of stars.

(From *The Economist*, 2018(7), pp. 69-70. by Daniel Franklin)

☞New Words

1. infrastructure [ˈɪnfrəstrʌktʃə(r)] *n.* the basic structure or features of a system or organization 基础设施
2. subscription [səbˈskrɪpʃn] *n.* a payment for consecutive issues of a newspaper or magazine for a given period of time 订阅
3. bazaar [bəˈzɑː(r)] *n.* a shop where a variety of goods are sold 集市；市场
4. forthcoming [ˌfɔːθˈkʌmɪŋ] *adj.* available when required or as promised 即将来临的
5. transaction [trænˈzækʃn] *n.* the act of transacting within or between groups (as carrying on commercial activities) 交易；事务；办理
6. vendor [ˈvendə(r)] *n.* someone who promotes or exchanges goods or services for money 卖主；小贩；供应商
7. scam [skæm] *vt.* to deprive of by deceit 欺诈；诓骗
8. bandwagon [ˈbændwægən] *n.* a popular trend that attracts growing support 流行，时尚
9. superficially [ˌsuːpəˈfɪʃəli] *adv.* in a superficial manner 表面地
10. reassure [ˌriːəˈʃʊə(r)] *vt.* to give or restore confidence in; to cause to feel sure or certain 使……安心，使消除疑虑
11. abduction [æbˈdʌkʃn] *n.* the criminal act of capturing and carrying away by force a family member 诱拐，绑架
12. contractor [kənˈtræktə(r)] *n.* someone (a person or firm) who contracts to build things 承包人；立契约者
13. resemble [rɪˈzembl] *vt.* to look like; to be similar or bear a likeness to 类似
14. incognito [ˌɪnkɒgˈniːtəʊ] *adv.* without revealing one's identity 隐姓埋名地
15. align [əˈlaɪn] *vt.* to be or come into adjustment 使结盟
16. deactivate [ˌdiːˈæktɪveɪt] *vt.* to remove from active military status or reassign 使无效
17. outsourcing [ˈaʊtsɔːsɪŋ] *n.* the contracting out of a business process to a third-party 外包；外购；外部采办
18. proliferation [prəˌlɪfəˈreɪʃn] *n.* a rapid increase in number 增殖，扩散
19. nudge [nʌdʒ] *vt.* to push into action by pestering or annoying gently 推进；用肘轻推

20. diner [ˈdaɪnə(r)] *n.* a person eating a meal (especially in a restaurant) 用餐者
21. thrive [θraɪv] *vi.* to grow stronger; to gain in wealth 繁荣，兴旺；茁壮成长
22. chagrin [ˈʃægrɪn] *n.* strong feelings of embarrassment 懊恼；委屈；气愤
23. discriminate [dɪˈskrɪmɪneɪt] *vt.* to recognize or perceive the difference 歧视；区别
24. disproportionately [ˌdɪsprəˈpɔːʃənətli] *adv.* out of proportion 不成比例地
25. conceal [kənˈsiːl] *vt.* to prevent from being seen or discovered 隐藏；隐瞒
26. underappreciate [ˌʌndərəˈpriːʃieɪt] *vt.* to look down upon 轻视
27. pseudonymous [suːˈdɒnɪməs] *adj.* bearing or identified by an assumed (often pen) name 匿名的；使用笔名的
28. disruptive [dɪsˈrʌptɪv] *adj.* characterized by unrest or disorder or insubordination 破坏的；分裂性的

☞Phrases and Expressions

1. jump on the bandwagon: to try to be in the swim 追赶趋势
2. align...with: to make correspond or harmonize 使……与……匹配
3. soak up: to use a great deal of money or other resources 占用；耗费
4. to one's chagrin: to cause to feel disappointed 令人懊恼的是

☞Notes

1. Uber: Uber Technologies Inc. is a ridesharing, taxi cab, food delivery, and bicycle-sharing transportation network company headquartered in San Francisco, California, with operations in 785 metropolitan areas worldwide.优步技术公司是一家集拼车、出租车、食品配送、自行车共享服务于一身的运输网络公司，总部设在加利福尼亚州的旧金山，其业务横跨世界范围内 785 个城市。
2. eBay: eBay Inc. is an American multinational e-commerce corporation based in San Jose, California that facilitates consumer-to-consumer and business-to-consumer sales through its website. 易趣公司是一家总部位于加利福尼亚州圣何塞的美国跨国电子商务公司，通过其网站为消费者对消费者和企业对消费者的销售提供便利。
3. Upwork: Upwork is a global freelancing platform where businesses and independent professionals connect and collaborate remotely. Upwork 是一个全球性的自由职业者平台，企业和独立专业人员可以远程连接和协作。
4. gig-economy: Gig-economy refers to an employment situation where the working arrangement is limited to a certain period of time based on the needs of the employing organization. 零工经济是指根据用人单位的需要，将工作时间安排限定在一定时间的就业形势。

☞Exercises

Vocabulary

Replace the underlined words or expressions in the following sentences with one of the best choices from the word bank.

A. incognito	B. proliferation	C. align with	D. to one's chagrin	E. infrastructure
F. superficially	G. disruptive	H. conceal	I. discriminate	J. resemble	K. nudge
L. soaks up	M. underappreciate	N. thrive	O. disproportionately	

1. A whole generation of people moving abroad is destructive to families, as well as the economy.
2. Defence takes up 40 percent of the budget.
3. It would not be difficult to pocket a weapon in a shoulder bag, he says.
4. Several professors openly urge students to go for PhD and look down upon bright students opting for the industry.
5. This is partly because there are unevenly more men in dangerous industries like construction and mining.
6. For our economy to prosper once again, we must remove these barriers to job creation.
7. We have joined forces to meet other challenges such as terrorism, piracy and nuclear propagation.
8. Marina O'Loughlin, London newspaper Metro's restaurant critic, has remained anonymous for the past 11 years.
9. It helps that, as the uprisings have spread, French interests happen to match their value.
10. Public products, in addition to basic facilities, include supplying system.

Unit 8 Finance and Accounting

✧ Text A

Taxes in the United States

In the United States, when you get your paycheck at the end of the first pay period at a new job, it's always interesting to see your net pay. Most of us expect more than we get. By the time you get your check, it has been cut up like a pizza, with several entities taking a piece of the pie. The entities that take money differ from person to person, company to company and state to state. However, almost every income earner has to pay federal income tax.

Taxes in Early America

Taxes have always left a sour taste in the mouth of American citizens. This national hatred for taxes dates back to the tax burden placed on the American colonies by Great Britain. Colonists were taxed for every consumer good, from tea and tobacco to legal documents. This "taxation without representation" led to many revolts, such as The Boston Tea Party, in which colonists dumped tea into the Boston Harbor rather than pay the tax on it.

Although the American colonists fought for independence from British rule and British taxes, once the United States government formed, its main source of revenue was derived from placing customs and excise taxes on the same items that were taxed by Great Britain. In 1812, in an effort to support an expensive war effort, the U.S. government imposed the first sales tax, which was placed on gold, silverware, jewelry and watches. In 1817, internal taxes were terminated and the government relied on tariffs to support itself. It wasn't until 1862 that the United State imposed the first national income tax.

To support the Union Army, Congress passed tax laws in both 1861 and 1862. The office of Commissioner of Internal Revenue was established by the Tax Act of 1862, which stated that the commissioner would have the power to levy and collect taxes. The office was also given the authority to seize property and income in order to enforce the tax laws. These powers remain pretty much the same today, although the IRS (Internal Revenue Service) will tell you that enforcement tactics have been toned down a bit.

In 1863, the federal government collected the first income tax. This graduated tax was similar to the income tax we pay today. Those who earned $ 600 to $ 10,000 per year paid at a rate of 3 percent. A higher rate was paid by those who earned in excess of $ 10,000. A fiat-rate tax was imposed in 1867. Five years later, in 1872, the national income tax was abolished altogether.

Inspired by the Populist Party's 1892 campaign, Congress passed the Income Tax Act of 1894. This act taxed 2 percent of personal income that was more than $ 4,000, which only affected the wealthiest citizens. The income tax was short-lived, as the U.S. Supreme Court struck it down only a year after it was passed. The justices wrote that, in their opinion, the income tax was unconstitutional because it failed to abide by a Constitutional guideline. This guideline required that any tax levied directly on individuals must be levied in proportion to a state's population.

In 1913, the income tax became a permanent part of the U.S. government. Congress avoided the constitutional roadblock mentioned above by passing a constitutional amendment. The 16th Amendment reads, "The Congress shall have the power to lay and collect taxes on incomes, from whatever source derived, without apportionment among the several states, and without regard to any census or enumeration."

Alternative: Flat Tax or National Sales Tax

Since the 16th amendment was passed in 1913, there has been no shortage of people proposing new tax systems since then. If you follow presidential campaigns, there are usually talks from some of the candidates on revising the tax system. Here's a quick look at two of these alternative tax plans.

The Flat Tax

We currently use a marginal tax system, also called a graduated tax, in which the percentage you pay in taxes varies based on your income. Under a flat tax system, you pay a flat rate on your income. In other words, there is a single tax bracket for all taxpayers. A common percentage thrown out for a flat-tax system is 17 percent. This is the rate proposed by former presidential candidate Steve Forbes and U.S. Representative Dick Armey.

Supporters of a flat tax system say that it would do away with the complicated tax code and tax forms. The flat tax would need only one form, about the size of a postcard and consisting of only 10 lines. You would merely add up wage, salary and pension income, subtract any personal allowances and pay 17 percent of your taxable income. Deductions and credits would be eliminated under this type of plan.

Critics of the flat tax say that it would favor the wealthy and could put a higher tax burden on those who make less money. Under Dick Armey's proposed flat tax, any family with a taxable income of less than $36,800 would pay no taxes. However, it would raise the taxes of some people who now may pay only 15 percent in taxes. The group who

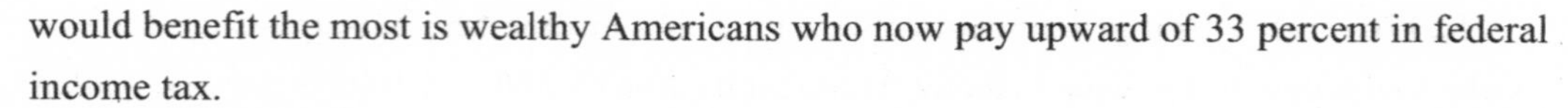

would benefit the most is wealthy Americans who now pay upward of 33 percent in federal income tax.

The National Sales Tax

Even more controversial than the flat tax is the idea of abolishing the federal income tax entirely by abolishing the 16th amendment. In place of an income tax, some propose the use of a national sales tax. Alan Keyes, who ran unsuccessfully for the 2000 Republican presidential nomination, is one of the biggest supporters of doing away with a federal income tax. He believes that we could finance the government through sales tax, tariffs and duties. Keyes has argued that a national sales tax would put more money back into the pockets of the consumers, letting them decide how to spend their own money. He says that the income tax should be replaced with the kind of taxes that people already pay. This plan would do away with the IRS and any need for a tax code.

Opponents have said that replacing the income tax with a national sales tax would put a heavy burden on the less wealthy, who buy a lot of products that would be taxed. They say that in order for a national sales tax to be fair, it would have to be applied to the purchase of stocks and bonds in addition to consumer goods. Another problem facing a national sales tax is that it would probably double the current taxes on consumer goods, and could force local and state governments to initiate or increase state income taxes.

(From *Income Tax in the United States.* 2010, pp. 45-47. by Frederic P. Miller)

☞New Words

1. entity [ˈentəti] *n.* something that exists separately from other things and has its own identity 实体；实际存在物
2. colonist [ˈkɒlənɪst] *n.* the people who start a colony or the people who are among the first to live in a particular colony 殖民者；殖民地居民
3. silverware [ˈsɪlvəweə(r)] *n.* objects that are made of or covered with silver, especially knives, forks, dishes, etc. that are used for eating and serving food 银餐具；银器，镀银器皿（尤指餐具）
4. permanent [ˈpɜːmənənt] *adj.* lasting for a long time or for all time in the future; existing all the time 永恒的，不变的，持久的
5. roadblock [ˈrəʊdblɒk] *n.* a barrier put across the road by the police or army so that they can stop and search vehicles 路障；<美>障碍
6. amendment [əˈmendmənt] *n.* the process of changing a law or a document 修正案；修改，修订
7. apportionment [əˈpɔːʃnmənt] *n.* the act of distributing by allotting or apportioning 土地分配；分摊，分派

8. census [ˈsensəs] *n.* the process of officially counting sth., especially a country's population, and recording various facts 人口普查，统计
9. enumeration [ɪˌnjuːməˈreɪʃn] *n.* the act of counting 计数，列举
10. alternative [ɔːlˈtɜːnətɪv] 1) *n.* a thing that you can choose to do or have out of two or more possibilities 可供选择的事物 2) *adj.* that can be used instead of sth. else 替代的；备选的
11. marginal [ˈmɑːdʒɪnl] *adj.* just barely adequate or within a lower limit 临界的，最低限度的；收入仅敷支出的
12. bracket [ˈbrækɪt] *vt.* to consider people or things to be similar or connected in some way 把……等同考虑；把……相提并论
13. complicated [ˈkɒmplɪkeɪtɪd] *adj.* made of many different things or parts that are connected; difficult to understand 结构复杂的；混乱的
14. pension [ˈpenʃn] *n.* an amount of money paid regularly by a government or company to sb. who is considered to be too old or too ill/sick to work 退休金，养老金
15. subtract [səbˈtrækt] *vt.* to take a number or an amount away from another number or amount 减去；扣除
16. allowance [əˈlaʊəns] *n.* an amount of money that is given to sb. regularly or for a particular purpose 津贴，补贴
17. nomination [ˌnɒmɪˈneɪʃn] *n.* the act of suggesting or choosing sb. as a candidate in an election, or for a job or an award; the fact of being suggested for this 提名；任命
18. tariff [ˈtærɪf] *n.* a tax that is paid on goods coming into or going out of a country 关税
19. opponent [əˈpəʊnənt] *n.* a person that you are playing or fighting against in a game, competition, argument, etc.对手；敌手；反对者
20. initiate [ɪˈnɪʃieɪt] *vt.* to make sth. begin 开始，发起

☞Phrases and Expressions

1. hatred for: to have a strong feeling of dislike for sb./sth. 对……的仇恨
2. derive from: to come or develop from sth.由……起源
3. abide by: to act in accordance with someone's rules, commands, or wishes 遵守；信守；忠于（某人）
4. without regard to: not to consider sth. 不考虑，不遵守
5. do away with: to remove it completely or put an end to sth. 废除，去掉

☞Notes

1. the Boston Tea Party: It is an incident in American history. It occurred on 16 December

1773. In order to protest about the British tax on tea, a group of Americans dressed as Indians went onto three British ships in Boston harbour and threw a number of boxes of tea into the sea. 波士顿倾茶事件：这是美国的历史事件。该事件发生于1773年12月16日。为了抗议英国征收的茶税，一群美国人乔装成印第安人进入停在波士顿海湾的3艘英国船只，把很多箱茶倒入了大海。

2. the IRS: Internal Revenue Service, the government department that is responsible for collecting most national taxes, for example, income tax. 美国国税局:负责征收大部分国税的政府部门，如所得税。
3. the Populist Party: founded in 2009, a minor political party that claims to advocate "classical liberalism" and a return to what they call "genuine" Constitutional government. 美国人民党：成立于2009年，是一个主张"古典自由主义"并回归他们所称的"真正"宪政的小政党。
4. Steve Forbes: As an American publishing executive, he was twice a candidate for the nomination of the Republican Party for President of the United States. Forbes is the Editor-in-Chief of *Forbes*, a business magazine. 史蒂夫·福布斯：美国出版业高管，曾两次当选美国共和党总统候选人。他是商业杂志《福布斯》的总编辑。
5. Alan Keyes：an American republican politician and the ambassador of the 16th Assistant Secretary of State for International Organization Affairs. 艾伦·凯斯：美国共和党政客，美国第16任国际组织事务助理国务卿。

☞Exercises

Ⅰ. Reading Comprehension

Answer the following questions according to the text.

1. What led to the independence of the North American colonies?
2. What is the purpose of imposing the sales tax by the government in 1812?
3. When did the United States government impose the first national income tax?
4. What does the flat tax intend to do?
5. Why do people oppose the National Sales Tax?

Ⅱ. Vocabulary

A. Replace the underlined words or expressions in the following sentences with one of the best choices from the word bank.

A. corruption	B. doubtful	C. impatient	D. endure	E. ignored	F. shrugged
G. laughable	H. applauding	I. likeable	J. obstacle	K. enjoyable	L. shiver
M. defeat	N. get the better of	O. uncertainty			

1. She's a <u>pleasant</u> young woman, who is always a very good company.
2. I'm afraid the reasons he gave me for not coming to the meeting were <u>silly</u> and unreasonable.
3. It is important to fight dishonest and illegal <u>behavior</u> by officials.
4. It's freezing out here. I'm shaking with <u>cold</u>.
5. You might find someone to help you in the office. but I'm <u>not sure.</u>
6. The attendant <u>moved his shoulders up</u> to suggest that he didn't know the answer.
7. Don't be so <u>annoyed</u> because the service is slow, and you won't miss your flight.
8. The sound of people <u>clapping</u> at the end of the Senator's speech lasted for five minutes.
9. The <u>snag</u> in this sort of anecdote is of course that one cannot distinguish between cause and effect.
10. I'm terribly squeamish. I can't <u>bear</u> gory films.

B. Fill in the blanks in each sentence by selecting the most suitable words.

1. ____ his advice, I would never have got the job.
   A. Except for B. Apart from C. But for D. As for
2. It's high time we ____ cutting down the rainforests.
   A. stopped B. had to stop C. shall stop D. stop
3. I heard the sound of the blind man ____ with his stick.
   A. clapping B. ticking C. touching D. tapping
4. Susan is very hardworking, but her pay is not ____ for her work.
   A. enough good B. good enough
   C. as good enough D. good as enough
5. There's no ____ in waiting for a bus: they don't run on public holidays.
   A. good B. point C. worth D. reason
6. Although the main characters in the novel are so true to life, they are entirely ____.
   A. imaginative B. imaginary C. imaginable D. imagined
7. He has always ____ strange hobbies like collecting bottle top and inventing secret codes.
   A. gone in for B. gone on
   C. gone with D. gone through with
8. It is an offence to show ____ against people of different races.
   A. distinction B. difference
   C. discrimination D. separation
9. ____ dull he may be, he is certainly a very successful top executive.
   A. Although B. Whatever C. As D. However
10. The party, ____ I was the guest of honour, was extremely enjoyable.
   A. at which B. by which C. to which D. for which

Ⅲ. Cloze

Choose an appropriate word from the four choices marked A, B, C and D for each blank in the passage.

For many people today, reading is no longer relaxation. To keep up their work they must read letters, reports, trade publications, interoffice communications, not to mention newspapers and magazines: a never-ending flood of words. In __1__ a job or advancing in one, the ability to read and comprehend quickly can mean the difference between success and failure. Yet the unfortunate fact is that most of us are poor readers. Most of us develop poor reading habits at an early age, and never get over them. The main deficiency __2__ in the actual stuff of language itself—words. Taken individually, words have __3__ meaning until they are strung together into phrases, sentences and paragraphs. __4__, however, the untrained reader does not read groups of words. He laboriously reads one word at a time, often regressing to reread words or passages. Regression, the tendency to look back over __5__ you have just read, is a common bad habit in reading. Another habit which slows down the speed of reading is vocalization—sounding each word either orally or mentally as one reads.

To overcome these bad habits, some reading clinics use a device called an __6__, which moves a bar (or curtain) down the page at a predetermined speed. The bar is set at a slightly faster rate __7__ the reader finds comfortable, in order to "stretch" him. The accelerator forces the reader to read fast, __8__ word-by-word reading, regression and sub-vocalization, practically impossible. At first, comprehension is sacrificed for speed. But when you learn to read ideas and concepts, you will not only read faster, __9__ your comprehension will improve. Many people have found their reading skills drastically improved after some training. Take Charlce Au, a business manager, for instance, his reading rate was a reasonably good 172 words a minute before the training, now it is an excellent 1,378 words a minute. He is delighted that how he can __10__ a lot more reading material in a short period of time.

1. A. applying	B. doing	C. offering	D. getting
2. A. lies	B. combines	C. touches	D. involves
3. A. some	B. a lot	C. little	D. dull
4. A. Fortunately	B. In fact	C. Logically	D. Unfortunately
5. A. what	B. which	C. that	D. if
6. A. accelerator	B. actor	C. amplifier	D. observer
7. A. then	B. as	C. beyond	D. than
8. A. enabling	B. leading	C. making	D. indicating
9. A. but	B. nor	C. or	D. for
10. A. master	B. get through	C. present	D.go over

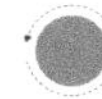

Ⅳ. Translation

Translate the following text into Chinese.

Many economists insisted that only a major restructuring of the world's debts could resolve the problem. In order to avert the danger of major defaults and also to restore stability to the economic system, Western banks and governments will have to ensure that they are imposing reasonable repayment terms on developing countries and that these terms will not provoke revolution. And they will have to strengthen the IMF (International Monetary Fund) as their ultimate safeguard. It is not enough for the IMF and the banks to provide a financial fire brigade moving from one debt crisis to another. The IMF must be given enough support to make longer-term loans in order to enable developing countries to revive their exports without enduring crippling hardships. Unless Western governments face up to this challenge, they may face political catastrophes that will require far more costly intervention.

Ⅴ. Writing

Write a composition entitled "Money". You should write at least 150 words according to the outline given below in Chinese.

1. 有些人认为金钱是生命中最重要的东西。
2. 但是金钱并不是万能的，也不能买到一切。
3. 来源正当的金钱才能使人幸福。

✧ Text B

The Alchemists of Finance

At least since 1823, when Byron's *Don Juan* described "Jew Rothschild, and his fellow Christian Baring" as the "true Lords of Europe", investment bankers have inspired awe, envy and, rightly or wrongly, a measure of disdain. Exactly 100 years ago, the undisputed patriarch of the modern industry, John Pierpont Morgan, stemmed the Panic of 1907, a financial crisis caused by unregulated trusts (the hedge funds of their day). Acting, in effect, as lender of last resort from his Wall Street office, he was briefly feted before Americans realized the danger of having such power vested in one man. Cartoonists then mercilessly mocked him. After his death in 1913, the Federal Reserve was set up.

The investment-banking industry was further constrained during the Depression of the 1930s, when Wall Street firms such as that founded by Morgan were split into commercial banks and securities houses. The later—today's investment banks—underwrite stocks and bonds and advise companies on mergers and acquisitions, rather than collect deposits and make loans. In the 1980s and 1990s, they developed a reputation for gluttonous excess. But a lot has changed since then.

Intensely private partnerships have become publicly traded companies. Commercial banks such as Citigroup and J. P. Morgan Chase have muscled back into investment banking. And European warhorses such as Deutsche Bank, UBS and Credit Suisse have joined the race for global supremacy. The bets, and the profits, have got bigger, though investment banks are trying to keep quiet about that, for several reasons.

First, they are under more scrutiny. Wall Street firms had their wings clipped by Eliot Spitzer, New York's former attorney-general, for plugging worthless shares during the dotcom era. Being publicly traded companies has tamed some egos, too. Star traders do not enjoy the same headroom on salaries (albeit very large salaries) as they did when they were partners in the business. At UBS, a Swiss bank which in 2000 moved into the American equity markets, merging with PaineWebber, a brokerage, "fiefs" are explicitly banned. Richard Fuld, boss of Lehman Brothers, a fast-growing Wall Street firm, imposed a "one-firm culture" when it was spun off from American Express in 1994. Now, says Scott Freidheim, a top executive, Mr. Fuld uses "culture" in speeches more than any other word except "the".

Meanwhile, another group has overtaken the investment banks in the excess stakes: their money-spinning clients in the private-equity and hedge-fund industries. Already they throw the biggest parties, do the boldest deals and launch the most celebrated initial public offerings. The IPO of part of Blackstone, a private-equity group, might well raise more money than Goldman Sachs's did in 1999, when even the company's doormen and drivers became extremely rich.

Yet when investment bankers discuss the fabulous fortunes accruing to these firms founders, they do so without envy. "Theirs is a truly pioneering role," says Anshu Jain, head of global markets at Deutsche Bank, one of the world's top trading banks. "Pioneers in any industry get a disproportionate share of the spoils."

Even if they are no longer the pioneers, the investment banks have played a crucial part in bringing about the extraordinary changes seen in the financial markets, starting in the 1980s and accelerating dramatically in the past five years. Technology and innovation have brought unprecedented breadth, depth and richness to financial instruments. According to McKinsey, a consultancy, the stock of shares and public and private debt securities held in America grew from 2.4 times GDP in 1995 to 3.3 times in 2004. In Europe, the increase was even more dramatic, albeit from a lower base. These figures do not include derivatives, notional amounts of which traded privately, or over-the-counter securities, which had soared to $370 trillion by last June, from $258 trillion less than two years earlier, according to the

Bank for International Settlements (BIS). Given such torrid growth, the markets are becoming increasingly vital to global financial stability.

There have been thrills and spills along the way. The stock market crash of 1987 and the seizing up of credit markets after Russia defaulted in 1998 both exposed huge flaws in the industry, forcing central banks to step in to prevent what they feared might be lasting damage to the real economy. Even so, regulators reckon that on balance the growth of markets has been a good thing, making the financial system safer than more traditional forms of bank lending. The trouble is that given the complexity of the new instruments and the range of clients and countries involved, they can never be absolutely sure that a monumental crisis is not brewing somewhere.

What worries both bankers and regulators is not so much the threat from hedge funds or private-equity groups but the implications for the financial system of a possible collapse of an investment bank (or large complex financial institution, as they clumsily call it). At a time when America's housing market has exposed the danger of overexcitement on Wall Street, it is worth exploring how these institutions are evolving, how they handle the risks attached to what they do, and how well those risks are spread around the financial system. That is what this survey sets out to do.

(From *The Economist*, 2007(3), pp. 93-94. by Anon David)

☞New Words

1. alchemist [ˈælkəmɪst] *n.* a person who studied alchemy 炼金术士
2. awe [ɔː] *n.* feelings of respect and slight fear; feelings of being very impressed by sth./sb. 敬畏；惊叹
3. disdain [dɪsˈdeɪn] *n.* the feeling that sb./sth. is not good enough to deserve your respect or attention 鄙视，轻蔑
4. patriarch [ˈpeɪtriɑːk] *n.* the male head of a family or community 家长，族长
5. stem [stem] *vt.* to stop sth. that is flowing from spreading or increasing 阻止；遏制
6. fete[feɪt] *vt.* to express respect and admiration for sb./sth.向……致敬
7. mock [mɒk] *vt.* to laugh at sb./sth. in an unkind way, especially by copying what they say or do 愚弄，嘲弄
8. underwrite [ˌʌndəˈraɪt] *vt.* to agree to buy shares that are not bought by the public when new shares are offered for sale 包销，承销
9. merger [ˈmɜːdʒə(r)] *n.* the act of joining two or more organizations or businesses into one （两个公司的）合并；联合体
10. gluttonous [ˈglʌtənəs] *adj.* given to excess in consumption of especially food or drink 贪吃的

11. scrutiny [ˈskruːtəni] *n.* careful and thorough examination 细看，细阅
12. fief [fiːf] *n.* an area of land, especially a rented area for which the payment is work, not money 封地，采邑
13. soar [sɔː(r)] *vi.* to rise rapidly 猛增，剧增
14. albeit [ˌɔːlˈbiːɪt] *conj.* although 虽然；即使
15. torrid [ˈtɒrɪd] *adj.* full of strong emotions 热烈的；狂热的
16. flaw [flɔː] *n.* a crack or fault in sth. that makes it less attractive or valuable 瑕疵，缺点
17. reckon [ˈrekən] *vi.* to expect, believe, or suppose 估计；猜想
18. clumsily [ˈklʌmzɪli] *adv.* in a clumsy manner 笨拙地

☞Phrases and Expressions

1. split…into: to divide sth. into two or more parts 分解为
2. keep quiet about: to refrain from divulging sensitive information 避免讨论……
3. merge with: to join together 和……联合
4. spin off: to turn a subsidiary into a new and separate company 分拆
5. bring about: to cause to happen, occur or exist 造成，实现
6. seize up: to become jammed 卡住；停顿

☞Notes

1. Rothschild: The Rothschild family is a European family of German Jewish origin that established European banking and finance houses from the late eighteenth century. It has been argued that during the 19th century, the family possessed by far the largest private fortune in the world, and by far the largest fortune in modern history. 罗斯柴尔德：罗斯柴尔德家族是一个起源于德国犹太人的欧洲家族，从18世纪末开始建立欧洲银行和金融机构。有人认为在19世纪，这个家族拥有当时世界上最大的私人财富，也是现代历史上最大的财富。
2. John Pierpont Morgan：John Pierpont Morgan was an American financier, banker and art collector who dominated corporate finance and industrial consolidation during his time. At the height of Morgan's career during the early 1900s, he and his partners had financial investments in many large corporations and were accused by critics of controlling the nation's high finance. He directed the banking coalition that stopped the Panic of 1907. He was the leading financier of the Progressive Era, and his dedication to efficiency and modernization helped transform American business. 约翰·皮尔庞特·摩根：他是美国金融家、银行家和艺术收藏家，他在自己的时代主导了公司金融和工业合并。20世纪初，在摩根事业的鼎盛时期，他和他的合伙人对许多大公司进行了金融投资，并

被批评家指责控制了国家的巨额融资，其领导的银行联盟阻止了 1907 年的恐慌。他是进步时代的金融领导家，并且他对效率和现代化的贡献助力了美国商业的转型。

3. Citigroup: 花旗集团。此外还有摩根大通（J. P. Morgan Chase）、德意志银行（Deutsche Bank）、瑞士瑞信银行（Credit Suisse）、高盛集团（Goldman Sachs）等。
4. Blackstone: The Blackstone Group L.P. is an alternative asset management and financial services company that specializes in private equity, real estate, and credit and marketable alternative investment strategies, as well as financial advisory services, such as mergers and acquisitions, restructurings and reorganizations, and private placements. 黑石集团：是一家另类资产管理和金融服务公司，专门从事私募股权、房地产、信贷和可销售另类投资策略，以及并购、重组和私募等金融咨询服务。
5. Bank for International Settlements (BIS): An intergovernmental organization of central banks which fosters international monetary and financial cooperation and serves as a bank for central banks. It is not accountable to any national government. The BIS carries out its work through subcommittees, the secretariats it hosts, and through its annual General Meeting of all members. It also provides banking services, but only to central banks, or to international organizations like itself. Based in Basel, Switzerland, the BIS was established by The Hague Agreement of 1930. 国际清算银行：中央银行的政府间组织，促进国际货币和金融合作，并作为中央银行的银行。它不对任何国家政府负责。国际清算银行通过其主办的小组委员会、秘书处以及所有成员的年度大会开展工作。它还提供银行服务，但仅限于中央银行或像其本身这样的国际组织。国际清算银行总部设在瑞士巴塞尔，根据 1930 年的《海牙协定》成立。

☞Exercise

Vocabulary

Replace the underlined words or expressions in the following sentences with one of the best choices from the word bank.

A. seize up	B. wasted	C. sensible	D. ordinary	E. rigid	F. keep in mind
G. stubborn	H. on balance	I. burst	J. peaceful	K. fixed on	L. allotment
M. bring down	N. splendid	O. in contrast			

1. The government has taken many measures to <u>reduce</u> the prices of oil.
2. A lot of people gathered at the airport waiting for the famous film star, and when he arrived, a <u>flurry</u> of excitement went around the crowd.
3. The result of many studies shows that they <u>dissipated</u> a lot of money when they carried out the so-called effective plans.
4. I was now in such a state of nervous resentment that I thought it <u>prudent</u> to check myself at

present from further demonstrations.

5. For more than a century, economics was largely concerned with problems other than growth, chiefly efficient allocation at the margin.
6. Even the most mundane thing can become objects of beauty in France's eye.
7. It was the earnest and dogmatic way in which she presented her political convictions that put everyone's back up.
8. You should bear in mind that powerful forces in the world are out to block the summit meeting.
9. I would say that, in general, it hasn't been a bad year.
10. Don't worry about his obstinate refusal to cooperate; he'll come to himself before long.

Keys to Exercises

Unit 1　Business and Trade

✧ Text A

Ⅰ. Reading Comprehension

1. In this article, business means the work or activities by which goods and services are provided and obtained for money payment.
2. Not all activities in which work is involved are classed as a business. Here is the test of whether or not an activity can be classed as a business: Is payment made for the goods supplied or service performed? If money payment is required, the activity is business.
3. People, goods, and money.
4. If just one phase of business, such as transportation, were to suspend operations, factories could not ship their products.
5. Because it furnishes the things we use, gives us useful work to do, offers opportunities for saving and investing, and aids the national defense.

Ⅱ. Vocabulary

A: 1. bounced back　2. typical of　3. derive from　4. interfere with　5. In some cases
6. spread to　7. is scheduled to　8. develops from　9. put forth　10. At the outset
B: 1-5 ADDDA　6-10 CBCAC

Ⅲ. Cloze

1-5 CAABC　6-10 DCBAC

Ⅳ. Translation

略

Ⅴ. Writing

略

✧ Text B

Vocabulary

1. cope with 2. dress up 3. bother with 4. plunged into 5. dated back to/dated from
6. reached for 7. condemned to 8. dependent on 9. comes down to 10. Chances are

Unit 2 Humanities and Arts

✧ Text A

Ⅰ. Reading Comprehension

1. He stayed around as a drop in for another 18 months or so before he really quit.
2. His biological mother was a young college graduate student when she gave birth to Steve Jobs as a single woman. She put up Steve Jobs for adoption, and felt strongly that her son should be adopted by college graduates and sent to college in the future.
3. His parents were working-class couple without a college degree. His mother never graduated from college and his father never graduated from high school. However, they did send their adopted son to college as they had promised, and this cost them nearly all their savings.
4. Because he had no idea what he wanted to do with his life and no idea how college was going to help him figure it out. Besides, the college was so expensive that all of his parents' savings were being spent on his tuition.
5. He thought it was one of the best decisions he had ever made as he looked back on his past. After he dropped out, he didn't have to take the required classes and he had the opportunity to take classes that interested him. Much of what he learned in this way proved priceless later on.

Ⅱ. Vocabulary

A: 1-5 I J G M K 6-10 C F H D B
B: 1-5 DBBCA 6-10 CCAAC

Ⅲ. Cloze

1-5 ACBCA 6-10 BDABA

Ⅳ. Translation

略

Ⅴ. Writing

略

✧ Text B

Vocabulary

1-5 B H F G N　6-10 M L C J E

Unit 3　Information and Technology

✧ Text A

Ⅰ. Reading Comprehension

1. Malaysia's WeChat users will be able to transfer money among themselves and make payments to offline merchants in ringgit.
2. Malaysia is a vibrant market. Technology-savvy Malaysians are embracing a digital lifestyle and to meet this shift, the payment experience has to evolve. Bringing WeChat Pay to Malaysia is their response to this.
3. Tencent is planning to build an open platform so that Internet service developers are able to provide more features for WeChat users and are developing value-added services for the application, especially in the mobile social contacts and online games sectors.
4. 开放题，依照个人理解表明观点即可。

Ⅱ. Vocabulary

A: 1-5 GBEIH　6-10 KMOFC
B: 1-5 DAACB　6-10 DADCC　11-16 ADBBDA

Ⅲ. Cloze

1-5 CABBA　6-10 CDCDC　11-15 DDBBA　16-20 DCBAC

Ⅳ. Translation

略

Ⅴ. Writing

略

✧ Text B

Vocabulary

1. rewarding 2. communicate 3. access 4. embarrassing 5. positive
6. commitment 7. virtual 8. benefit 9. minimum 10. opportunities

Unit 4 Aeronautics and Space

✧ Text A

Ⅰ. Reading Comprehension

1. That helped make it possible to find the International Space Station just by looking up.
2. The space station was carrying six crew members.
3. He is a pilot, an officer in the Russian Air Force and a medical doctor who was born in Crimea.
4. Three others joined them after leaving the Baikonur Cosmodrome on March 25.
5. The Maryland sky watchers found the ISS by looking to the northwest and watching it move southeast.

Ⅱ. Vocabulary

A: 1-5 ADHEJ 6-10 OLMNB
B: 1-5 CAABA 6-10 BCDAB

Ⅲ. Cloze

1-5 CDCBB 6-10 DAAAB 11-15 BDDAD 16-20 ACBBB

Ⅳ. Translation

略

Ⅴ. Writing

略

✧ Text B

Vocabulary

1-5 OKGJC 6-10 NMIHD

Unit 5 Material Engineering

✧ Text A

Ⅰ. Reading Comprehension

1. With a hand clad in chain mail, the underlying skeleton and thin, meshlike skin, it looks like something out of medieval times.
2. It has troubles in additive manufacturing, the fit and finish of its materials, and in building objects out of multiple kinds of materials. And they cannot integrate electronics without frying the circuits.
3. No, it doesn't. Because it has taken two extreme paths since it was invented.
4. Printing in 3-D could replace certain conventional mass-production processes such as casting, molding and machining by 2030, especially in the case of short production runs or manufacturers aiming for more customized products.

Ⅱ. Vocabulary

A: 1-5 DCALJ 6-10 IGFBK
B: 1-5 DBACB 6-10 CCABD

Ⅲ. Cloze

1-5 BDDAB 6-10 DBBCD 11-15AACAB

Ⅳ. Translation

略

Ⅴ. Writing

略

✧ Text B

Vocabulary

1-5 BCBCD 6-10 BCADB

Unit 6 Mechanical Engineering

✧ Text A

Ⅰ. Reading Comprehension

1. The knuckle joint assembly consists of the following major components: Single eye, Double eye or fork, Knuckle pin.
2. a. Tie rod joint of a roof truss. b. Tension link in bridge structure. c. Link of roller chain. d. Tie rod joint of the jib crane. e. The knuckle joint is also used in the tractor.
3. Four. a. To tighten the members of the roof truss. b. Used to connect link in a mechanism to transfer motion. c. Used between the two railways wagon and bogies. d. To tighten the cable or stay ropes of electric distribution poles.
4. Tension joints and shear joints
5. a. Quick assembly and disassembly is possible. b. It can take tensile as well as compressive force.

Ⅱ. Vocabulary

A: 1. the norm 2. although 3. plentiful 4. investigated 5. adapted
6. Apart from 7. expectations 8. energetic 9. greedy 10. doubtful

B: 1-5 AAACB 6-10 DDBDB

Ⅲ. Cloze

1-5 BDABC 6-10 BABDB

Ⅳ. Translation

略

Ⅴ. Writing

略

✧ Text B

Vocabulary

1. put in 2. make the most of 3. put down 4. lead to 5. get her hands on
6. put away our picnic things 7. to stick to 8. time after time

9. has cut down on smoking　　10. counts for much more

Unit 7　Economics and Management

✧ Text A

Ⅰ. Reading Comprehension

1. There could be a 300m-400m horde of potential frappuccino-sippers, Fiesta-drivers and globe-trotters.
2. The top 1% of Indian adults, a rich enclave of 8m inhabitants making at least $20,000 a year; The next 9% is in the middle of the global wealth pack; The next 40% of India's population neatly mirrors its combined South Asian poor neighbours; The remaining half-billion or so are on a par with the most destitute bits of Africa.
3. The top 1% of earners pocketed nearly a third of all the extra income generated by economic growth between 1980 and 2014.
4. Peddling wares using regional languages preferred by hundreds of millions of Indians.
5. Goods and services that offer stonking value.

Ⅱ. Vocabulary

A: 1-5 KGBMC　6-10 LDJAO
B: 1-5 CDBDB　6-10 BABAB

Ⅲ. Cloze

1-5 BBDBB　6-10 ABDBC

Ⅳ. Translation

略

Ⅴ. Writing

略

✧ Text B

Vocabulary

1-5 GLHMO　6-10 NBACE

Unit 8 Finance and Accounting

✧ Text A

Ⅰ. Reading Comprehension

1. Colonists were taxed for every consumer good, from tea and tobacco to legal documents, and the tax burden placed on the American colonies led to independence.
2. In 1812, in an effort to support an expensive war effort, the U.S. government imposed the first sales tax, which was placed on gold, silverware, jewelry and watches.
3. It wasn't until 1862 that the United State imposed the first national income tax.
4. The flat tax intends to do away with the complicated tax code and tax form in the United States.
5. People who oppose the National Sales Tax maintain that it would be unfair to the less wealthy because it is only applied to the purchase of consumer goods.

Ⅱ. Vocabulary

A: 1-5 IGALB 6-10 FCHJD
B: 1-5 CADBB 6-10 BACDA

Ⅲ. Cloze

1-5 DACDA 6-10 ADCAB

Ⅳ. Translation

略

Ⅴ. Writing

略

✧ Text B

Vocabulary

1-5 MIBCL 6-10 DEFHG

Main References

[美]阿米里奥，西蒙. 2002. IT 帝国——苹果公司转型中的管理风波. 孟详成译. 北京：中国建材工业出版社.

鲍文. 2009. 国际商务英语学科论. 北京：国防工业出版社.

蔡越坤. 2016. iCar 才是未来！苹果被曝电动车投入已远超 iPhone. 2016-5-26. https://business.sohu.com/20160526/n451525952.shtml [2019-9-26].

顾舜若. 2010.《杀死一只反舌鸟》中的信仰格局. 当代外国文学, (1): 168-170.

姜虹宇. 2016. 传说中的苹果 iCar 到底在哪里？2016-8-2. http://digi.163.com/16/0802/09/BTF3G14D00162Q5T_all.html [2019-9-2].

凯尔纳. 2010. 恩斯特·布洛赫：乌托邦与意识形态批判. 王峰译. 马克思主义美学研究, (1): 64-77.

李恩运. 2008. 演讲语篇中的语气及情态系统的人际功能分析. 科技信息, (18): 478, 481.

李红梅. 2013. 浅析《杀死一只反舌鸟》的种族歧视主题. 作家, (6X): 40-41.

李运兴. 2003. 英汉语篇翻译. 北京：清华大学出版社.

彭宣维. 2000. 英汉语篇综合对比. 上海：上海外语教育出版社.

施平. 2006. 先进制造技术(*Advanced Manufacturing Technology*). 哈尔滨：哈尔滨工业大学出版社.

施平. 2014. 机电工程专业英语. 9 版. 哈尔滨：哈尔滨工业大学出版社.

施平. 2015. 机械工程专业英语教程. 4 版. 北京：电子工业出版社.

[美]扬，西蒙. 2010. 活着就为改变世界：史蒂夫·乔布斯传. 蒋永军译. 北京：中信出版社.

张佐成. 2008. 商务英语的理论与实践研究. 北京：对外经济贸易大学出版社.

Anand, M. K. 2016. Reforming fossil fuel prices in India: Dilemma of a developing economy. *Energy Policy*, 92: 139-150.

Bargiela-Chiappini, F. & Nickerson, C. 2003. Intercultural Business Communication: A rich field of studies. *Journal of Intercultural Studies*, 24(1): 3-15.

Brown, T. M., Halliday, M. A. K., McIntosh, A., et al. 1964. *The Linguistic Sciences and Language Teaching*. London: Longman.

Bourne, D. 2013. My boss the robot. *Scientific American*, 308 (5): 38-41.

Brieger, N. 1997. *Teaching Business English Handbook*. England: York Associates.

Chai, H. 2018. Tencent's Pony Ma launches guidebook for Bay Area future. 2018-7-24.

http://www.chinadaily.com.cn/a/201807/24/WS5b569c23a31031a351e8fc9e.html [2019-9-24].

Cross, N. 2008. *Engineering Design Methods: Strategies for Product Design.* Trenton: John Wiley & Sons.

Dudley-Evans, T. & St John, M. J. 1998. *Developments in English for Specific Purposes: A Multi-Disciplinary Approach*. Cambridge: Cambridge University Press.

Falebita, O. & Koul, S. 2018. From developing to sustainable economy: A comparative assessment of India and Nigeria. *Environmental Development*, 25: 130-137.

Greenemeier, L. 2013. To print the impossible. *Scientific American*, 308 (5): 44-47.

Isaacson, W. 2012. The real leadership: Lessons of Steve. *Harvard Business Review*, 18: 5.

Jobs, S. 2005-6-12. 'You've got to find what you love,' Jobs says. http://news.stanford.edu/news/ 2005/june15/jobs-061505.html[2018-10-8].

Lucey, P. G., Norman, J. & Crites, S., et al. 2014. A large spectral survey of small lunar craters: Implications for the composition of the lunar mantle. *American Mineralogist*, 99(11): 2251-2257.

Schmid, K. 2006. *Manufacturing Engineering and Technology*. New York: Pearson Education.